LONDON MATHEMATICAL SOCIETY STUDENT TEXTS

Managing editor: Dr C.M. Series, Mathematics Institute
University of Warwick, Coventry CV4 7AL, United Kingdom

London Mathematical Society Student Texts 5

An Introduction
to General Relativity

L. P. Hughston
Merrill Lynch International, London
and King's College, London

K. P. Tod
University of Oxford

CAMBRIDGE
UNIVERSITY PRESS

Published by the Press Syndicate of the University of Cambridge
The Pitt Building, Trumpington Street, Cambridge CB2 1RP
40 West 20th Street, New York, NY 10011-4211, USA
10 Stamford Road, Oakleigh, Melbourne 3166, Australia

© Cambridge University Press 1990

First published 1990
Reprinted 1992, 1994

Library of Congress cataloging in publication data available

A catalogue record for this book is available from the British Library

ISBN   0 521 33943 X   paperback

Transferred to digital printing 2000

Contents

Preface

Omnia profecto cum se coelestibus rebus referet ad humanas, excelsius magnificentiusque, et dicus et sentiet. (*The contemplation of celestial things will make a man both speak and think more sublimely and magnificently when he descends to human affairs.*)

—Cicero

IT IS INEVITABLE that with the passage of time Einstein's general relativity theory, his theory of gravitation, will be taught more frequently at an undergraduate level. It is a difficult theory—but just as some athletic records fifty years ago might have been deemed nearly impossible to achieve, and today will be surpassed regularly by well-trained university sportsmen, likewise Einstein's theory, now over seventy-five years since creation, is after a lengthy gestation making its way into the world of undergraduate mathematics and physics courses, and finding a more or less permanent place in the syllabus of such courses. The theory can now be considered both an accessible and a worthy, serious object of study by mathematics and physics students alike who may be rather above average in their aptitude for these subjects, but who are not necessarily proposing, say, to embark on an academic career in the mathematical sciences. This is an excellent state of affairs, and can be regarded, perhaps, as yet another aspect of the overall success of the theory.

But the study of general relativity at an undergraduate level does present some special problems. First of all, the content of the course must be reasonably well circumscribed. At a graduate or research level treatment it may be necessary and even desirable for the course to veer off asymptotically into more and more difficult and obscure material, eventually reaching the 'cutting edge' of the subject (the edge where the theory no longer cuts). For an undergraduate course this will not do—and this will mean some material has to be omitted; but that is not a serious worry: what *is* required (and this goes especially for the presentation in lectures) is that subtle blend of seriousness and stimulation that cannot really be prescribed or explained, but is as rare as it is easily recognized. Whether we have fulfilled *this* requirement very satisfactorily is doubtful; but we *have* succeeded in omitting some material.

Another function that an undergraduate course must satisfy is that is should be *examinable*. This means slightly less emphasis on the sort of lengthy calculations and verifications that are typically put forth as problems in the 'trade' books (though such problems are in the right context useful and important) and more emphasis on the slightly shorter type of problem that requires some *thinking* for its solution—problems that, as G.H. Hardy might have said, show a bit of *spin*. We cannot claim that our problems are being bowled so artfully, or even that that's always what's intended;

but it *can* be pointed out that a number of the problems appearing at the ends of chapters *have* been set on past papers of undergraduate examinations at Oxford, and that these, and other problems set in the same spirit, may be useful not only for the lofty purpose of enriching one's comprehension of a noble subject, but also for the mundane but very important matter of proving to the rest of the world that one's comprehension has indeed been enriched!

A number of our colleagues have helped us in various ways in the preparation of this material—either by providing us with problems or ideas for problems, or by reading portions of the original lecture notes on which the course is based and offering criticism and useful feedback, or pointing out errors; and we would particularly like to thank David Bernstein, Tom Hurd, Lionel Mason, Tristan Needham, David Samuel, and Nick Woodhouse for this. Roger Penrose has in his publications and lectures suggested a number of points of approach and presentation that we have used or adapted, for which we offer here summarily our acknowledgements and thanks—for indeed much of the subject as it presently stands bears the imprint of his significant influence. And for the mathematical typesetting we would like to thank Jian Peng of Oxford University Computing Laboratory.

L.P.Hughston
Robert Fleming & Co. Limited
25 Copthall Avenue
London EC2R 7DR
United Kingdom

K.P.Tod
The Mathematical Institute
Oxford OX1 3LB
United Kingdom

1   Introduction

1.1   Space, time, and gravitation

GENERAL RELATIVITY is Einstein's theory of gravitation. It is not only a theory of gravity: it is a theory of the structure of space and time, and hence a theory of the dynamics of the universe in its entirety. The theory is a vast edifice of pure geometry, indisputably elegant, and of great mathematical interest.

When general relativity emerged in its definitive form in November 1915, and became more widely known the following year with the publication of Einstein's famous exposé *Die Grundlage der allgemeinen Relativitätstheorie* in *Annalen der Physik*, the notions it propounded constituted a unique, revolutionary contribution to the progress of science. The story of its rapid, dramatic confirmation by the bending-of-light measurements associated with the eclipse of 1919 is thrilling part of the scientific history. The theory was quickly accepted as physically correct—but at the same time acquired a reputation for formidable mathematical complexity. So much so that it is said that when an American newspaper reporter asked Sir Arthur Eddington (the celebrated astronomer who had led the successful solar eclipse expedition) whether it was true that only three people in the world really understood general relativity, Eddington swiftly replied, "Ah, yes—but who's the third?"

The revolutionary character of Einstein's gravitational theory lies in the change of attitude towards space and time that it demands from us. Following Einstein's extraordinary 1905 paper on special relativity (*Zur Elektrodynamik bewegter Körper*, in *Annalen der Physik*) a major step forward was taken by the mathematician Hermann Minkowski (1864-1909) who recognized that the correct way to view special relativity, and in particular the Lorentz transformation, was in terms of a single entity *space-time*, rather than a mere jumbling up of space and time coordinates.

His famous 1908 address *Space and Time* opens theatrically with the following words: "The views of space and time that I wish to lay before you have sprung from the soil of experimental physics, and therein lies their strength. They are radical. Henceforth space by itself and time by itself, are doomed to fade away into mere shadows, and only a kind of union of the two will preserve an independent reality." And how right he was. It was in Minkowski's work that many of the geometrical ideas so important to a correct, thorough understanding of relativity were first introduced—particle world-lines, space-like and time-like vectors, the forward and backward null cones—ideas that all carry over also into general relativity. Minkowski space (as the flat Lorentzian space-time manifold of special relativity is now called) was seen to be the proper arena for the description of special relativistic physical phenonena—a point of view that Einstein himself was quickly to embrace. "We are compelled to admit," writes Minkowski in the same 1908 address, "that it is only in four dimensions that the relations here taken under consideration reveal their inner being in full simplicity, and that on a three dimensional space forced upon us a *priori* they cast a very complicated projection."

1.2   The dynamics of the universe in its entirety

But general relativity goes much further, and incorporates the gravitational field into the structure of space-time itself. Since gravitational fields can vary from place to place, this means that space-time also must vary in some way from place to place. The mathematical framework that deals with geometries that vary from point to point is called *differential geometry*; and it is a particular species of differential geometry called *Riemannian geometry*—named after Bernhard Riemann (1826-1866) who among his many mathematical achievements founded the general theory of higher dimensional curved spaces—that offers the analytical basis for a description of the gravitational field. Einstein himself was to remark (in *The Meaning of Relativity*) that "the mathematical knowledge that has made it possible to establish the general theory of relativity we owe to the geometrical investigations of Gauss and Riemann."

And thus we are left to marvel that Einstein was led to such a refined, abstract branch of pure geometry for his relativistic theory of gravitation. " It is my conviction," he writes in his 1933 Herbert Spencer lecture at Oxford, "that pure mathematical construction enables us to discover the concepts, and the laws connecting them, that give us the key to the understanding of the phenomena of Nature." It was easy, perhaps, for Einstein to say this in 1933. By that time he was recognized throughout the world as a genius. His theories had transformed the shape of physical science. And yet another accolade has been bestowed upon him in the years immediately preceding—the American astronomer Edwin Hubble (1889-1953) had announced in 1929 his discovery that the universe was expanding!—that remote galaxies showed a red shift systematically correlated with their distance. This observation was very much in accord with the pattern of results suggested by general relativity, and opened the door to yet another new branch of physics: relativistic cosmology.

1.3   What is so special about general relativity?

The road to special relativity had been swift and straight, with most of the essentials accomplished in Einstein's first article on the subject, his 1905 paper. The famous $E = mc^2$ formula follows shortly thereafter in a brief note entitled *Ist die Trägheit eines Körpers von seinem Energiegehalt abhängig?* (Does the inertia of a body depend upon its energy-content?) which ends with the speculation that "It is not impossible that with bodies whose energy-content is variable to a high degree (e.g. with radium salts) the theory may be successfully put to the test." Capped with Minkowski's mathematics, the theory was then set for a successful launch onto the high seas of twentieth century physics—and is still very much afloat.

So much for Albert Einstein (1879-1955) at age twenty-six: the genesis of general relativity, however, was a far less straightforward matter, and took the better part of a decade. One is reminded of the way in which some musical pieces seem to have sprung, as it were, fully composed from the musician's head—one gets this impression, for example, in many of the writings of Bach; whereas other pieces are clearly arrived at only after extensive revisions, with ideas being gradually assembled in the course of a tortured creative process spread over a period of some time—in this case Beethoven comes to mind as possibly the best example, and one also thinks particularly of Mahler. If special relativity belongs to the first category of composition, then general relativity certainly falls into the second.

Nevertheless, despite its complex origins, a Beethoven symphony or quartet does have a certain finality to it—a certain undeniable perfection; and much the same can be said of Einstein's gravitational theory. It has that rare quality about it that excites all of one's attentions in a physical theory: it has an air of permanence.

And it is, perhaps, this aspect of Einstein's theory that makes it (quite apart from its necessary interest to professional physicists, as a key component to our present understanding of nature) a subject worthy of intellectual enquiry by students who, after coming to understand it, will not in any ordinary sense have any practical use for it. It is a work of art.

1.4 The mercurial matter of Mercury

In December 1907 Einstein wrote to his friend Conrad Habicht (1876-1958) that he was "...busy working on relativity theory in connection with the law of gravitation, with which I hope to account for the still unexplained secular changes in the perihelion motion of the planet Mercury—so far it doesn't seem to work." Einstein, Habicht, and another friend Maurice Solovine (1875-1958) had known one another in Bern (where Einstein had taken up his job at the patent office in 1902) and had met regularly under the auspices of the *Olympian Academy* (founded by and comprising just the three of them) to discuss and debate philosophical, scientific, and literary matters. They read together from the works of Plato, Sophocles, Cervantes, Hume, Spinoza, Racine, Dickens, Mach, and Poincaré, amongst other authors. How exciting those evenings must have been! What else might they have read? (One is reminded somehow of Oscar Wilde's remark, "I have made an important discovery—that alcohol, taken in sufficient quantities, produces all the effects of intoxication.") Einstein evidently felt at ease with Habicht to discuss his ambitions and frustrations. And it was eight years later in 1915 that Einstein was able in a letter to the physicist Paul Ehrenfest (1880-1933) to report that "for a few days, I was beside myself with joyous excitement" over the correct explanation of Mercury's orbit, which he had recently obtained.

1.5   An idée fixe

But why this apparent obsession with the misbehaviour of Mercury's orbit? Why dwell on this little detail? The problem with Mercury had been known since the middle of the nineteenth century. According to Newton's theory, and as first hypothesized by Kepler, the orbit of an ideal planet is a perfect ellipse, with the Sun located at one of the foci, as illustrated below.

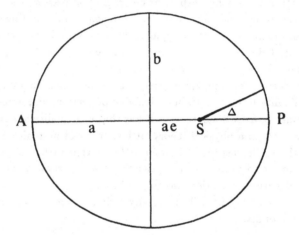

Figure 1.1.   The shift of Mercury's perihelion. With each revolution the axis of the ellipse moves through a small angle Δ in the direction of revolution.

The semi-major axis a and the semi-minor axis b are related by $b^2 = a^2(1 - e^2)$, where e is the eccentricity. The sun is offset from the center of the ellipse by a distance ae, and the equation of the orbit is given by

$$r = \frac{a(1 - e^2)}{1 + e \cos \theta}$$

where r is the distance between the sun and the planet, and θ is the angle between the line joining the sun and the planet, and the semi-major axis through S. Clearly where $\theta = 0$ the planet is at its *perihelion* (point of closest approach) with $r = a(1 - e)$; whereas when $\theta = \pi$ the planet is most distant, at its aphelion, with $r = a(1 + e)$.

In reality planetary orbits are not perfect ellipses, owing primarily to the perturbing effects of other planets. This was of course well-appreciated by Newton, who in *De Motu* (version IIIB) observes that "...the planets neither move exactly in an ellipse, nor revolve twice in the same orbit—there are as many orbits to a planet as it has revolutions—and the orbit of any one planet depends on the combined motion of all of the planets, not to mention the action of all these on each other. But to consider simultaneously all the causes of motion and to define these motions by exact laws allowing of convenient calculation exceeds, unless I am mistaken, the force of the entire human intellect."

Fortunately, the perturbing effects are relatively small, and one can treat the orbits as approximately elliptical, studying the deviation from perfect ellipticity, as induced by effects such as those mentioned by Newton. Notable among these is the effect of a small rotation in the axis of the ellipse, which can be measured by the angle Δ by which the perihelion shifts, per revolution, from its previous position. For Mercury, planetary influences result in a perihelion shift of roughly 500″ (seconds of arc) per century. Since Mercury's period is about one quarter that of the Earth, this works out to about 1.25″ per orbit—not very much! Around 260,000 years are required for the precession to go all the way around. But it does happen.

The anomaly in Mercury's orbit was discovered by the French astronomer Urbain Jean Joseph Le Verrier (1811-1877) in 1859. He showed that there was a discrepancy between observation and theory—by a figure which (according to present-day measurements) amounts to an excess motion in the perihelion shift of about 43″ per century. This was the 'still unexplained change' in Mercury's orbit that had caught Einstein's attention in 1907.

1.6   Beside himself with joy

The history of attempts to explain this phenomenon is an elaborate affair, and makes a very interesting chapter in the history of astronomy. In essence, either a mysterious new planet, or some form of hidden quasi-planetary material, had to be present—or the laws of gravity needed to be modified.

On the latter point it is indeed straightforward enough to 'induce' a systematic perihelion shift by means of a slight modification of Newton's laws: Newton himself had noted, for example, that if the gravitational force obeyed not an inverse square law but rather, say, a modified force-law of the form

$$F = \frac{\alpha r^m - \beta r^n}{r^3}$$

where α, β, m, n are constants, then the angle θ between successive perihelia ($\theta = 2\pi + \Delta$) is given by

$$\theta = 2\pi \left(\frac{\alpha - \beta}{m\alpha - n\beta} \right)^{1/2}$$

Thus in particular if $F = \alpha r^{m-3}$ we have $\theta = 2\pi m^{-1/2}$. By today's way of thinking (which as a consequence of Einstein's scientific work has become much more Platonic) such a modification of Newton's theory strikes us as rather vulgar—but a hundred years ago the approach was taken seriously as a possible explanation of Mercury's anomaly. It doesn't work.

Einstein, however, was able in his theory to deduce a very elegant formula for the perihelion effect, given by

$$\Delta = \frac{24\pi^3 a^2}{T^2 c^2 (1 - e^2)}$$

where T is the period of the orbit, and c is the speed of light. No wonder, given the accuracy with which it accounts for the observed orbits, that Einstein was beside himself with joy at the discovery of this relation. One can imagine the profound shock it must have given him to have encountered such a vivid confirmation of his ideas—a confirmation of the sublime relations holding between the abstractions of the space-time continuum, and something so down to earth as the science of the solar system.

1.7   Rudis indigestaque moles

This may give us an intimation as to why the theory has been lifted to such preeminent esteem by the cognoscente of successive generations. By comparison, the scope of other physical theories, indeed much of science as a whole, takes on the character of 'a rough and confused mass.'

General relativity is a theory of some complexity, and it does involve a good deal of fairly difficult mathematics. Nevertheless it is possible—providing one is willing to take a number of details on faith—to present an overview of the theory, reducing it to its most basic mathematical elements.

Space-time, according to Einstein's theory, is a *four-dimensional manifold* (the 'space-time continuum'). The manifold looks locally like a piece of R^4, but there are two important distinctions: the various 'pieces' do not necessarily fit together to form a *global* R^4 (what they do form is typically something more complicated); and even locally the geometry is not Euclidean, nor even flat (like the Lorentzian geometry of special relativity).

The space-time is covered by a series of coordinate patches U_i, and in each coordinate patch we have a set of coordinates x^a ($a = 0, 1, 2, 3$). The basic, underlying geometry of the manifold (its *differentiable structure*) is determined by the relations holding between systems of coordinates in overlapping patches.

The mathematical tool used for studying manifolds is called *tensor calculus*. A tensor is a sort of a many-index analogue of a vector. Differentiation of tensors is a intricate matter since the value of the derivative of a tensor can apparently depend (in a coordinate overlap region) on which set of coordinates is used to perform the differentiation. This situation is remedied by the introduction in each coordinate patch of a special three-index array of functions denoted Γ^a_{bc} called the 'connection'. The connection is required to transform in a particular way in coordinate transition (i.e. overlap) regions. The correct derivative of a tensor is then taken by means of a slightly complicated operation that involves systematic use of these special connection symbols. The resulting process—called 'covariant differentiation' has a multitude of natural, compelling features. (The covariant derivative of a tensor A^{bc} is denoted $\nabla_a A^{bc}$ to distinguish it from the array of partial derivatives $\partial_a A^{bc}$ where $\partial_a = \partial/\partial x^a$. More explicitly, the covariant derivative $\nabla_a A^{bc}$ is given by an expression of the form $\partial_a A^{bc} + \Gamma^b_{ap} A^{pc} + \Gamma^c_{aq} A^{bq}$.) When the operations of calculus in this way become well-defined we say that the space-time has the structure of a *differentiable manifold with connection*.

1.8   The metric tensor

But in what sense is the manifold a *space-time* manifold? In special relativity the metrical properties of space and time are determined by a 'Lorentzian' metric with a pseudo-Euclidean signature, given by

$$ds^2 = dt^2 - dx^2 - dy^2 - dz^2.$$

In the case of a material particle the infinitesimal interval ds represents the change in 'proper time' (i.e. time as measured by its own natural clock) experienced by the particle when in undergoes a displacement in space-time given by (dt, dx, dy, dz). Units are chosen such that the speed of light is one. Note that if dx, dy, and dz vanish, then the proper time s agrees with the coordinate t. Thus t can be interpreted as the time measured by an observer at rest at the origin in this system of coordinates. But if dx is greater than zero, say, then ds must be less than dt. So we see that while an observer at the origin measures an interval of time dt, a *moving* body instead measures the interval given by $ds^2 = dt^2 - dx^2$, or more explicitly $ds = (1 - v^2)^{\frac{1}{2}} dt$ where $v = dx/dt$ is the velocity of the moving observer relative to the origin. And thus time seems to be going more 'slowly' for the moving particle.

The infinitesimal interval of special relativity can be written more compactly by use of an index notation in the form

$$ds^2 = \eta_{ab} dx^a dx^b$$

where dx^a ($a = 0, 1, 2, 3$) is the space-time displacement, and η_{ab} is the flat metric of special relativity with diagonal components $(1, -1, -1, -1)$. Summation is implied over the repeated indices. In general relativity the idea is that the geometry of space-time varies from point to point—and this is represented by allowing the metric to be described by a tensor field g_{ab} that varies over the space-time. The infinitesimal interval is then given by

$$ds^2 = g_{ab} dx^a dx^b,$$

where g_{ab} is a four-by-four symmetric, non-degenerate matrix. It was Einstein's key recognition that the gravitational field could be embodied in the specification of the space-time manifold M and its 'curved' Lorentzian metric g_{ab}. The idea that M is a *space-time* is implicit in the requirement that g_{ab} should have signature $(+, -, -, -)$; i.e. that it should have three negative eigenvalues, and one positive eigenvalue.

One immediate consequence of the formula $ds^2 = g_{ab}dx^a dx^b$ is that the proper time experienced by a particle depends on the nature of the gravitational field through which it may be passing. This leads to a gravitational *time dilation* effect, which is one of the important features of the theory—when light is emitted in the neighbourhood of a strong gravitational field (e.g. near the Sun) it is seen to be *red-shifted* when received in the vicinity of a weaker field (e.g. at the Earth's surface).

1.9   The Levi-Civita connection

The flat space-time of special relativity is called *Minkowski space*, and g_{ab} can be regarded as 'fixed' throughout the manifold. But in a curved space-time g_{ab} varies from point to point in an *essential* way.

Therefore to set up a workable physical theory one needs a relation between the metrical properties of the space-time (as determined by g_{ab}) and the operations of tensor calculus (as determined by the connection Γ^c_{ab}). This is established by a powerful result known as the *fundamental theorem of Riemannian geometry*. According to this theorem the space-time metric determines the associated connection Γ^c_{ab} according to a remarkable formula, given by

$$\Gamma^a_{bc} = \frac{1}{2}g^{ad}(\partial_b g_{cd} + \partial_c g_{bd} - \partial_d g_{bc})$$

where g^{ab} is the inverse of g_{ab} (so $g^{ab}g_{bc} = \delta^a_c$), and ∂_a again denotes $\partial/\partial x^a$. The connection thus determined is called the *Levi-Civita* connection, and the corresponding covariant derivative ∇_a has the important property that when applied to the metric tensor it gives the result zero: $\nabla_a g_{bc} = 0$, i.e. the metric is 'covariantly constant'. The point of the theorem is that given g_{ab} the connection is determined uniquely by this property.

1.10   The field equations

Sitting at the apex of the theory are Einstein's equations for the gravitational field. These are the equations that relate g_{ab} to the local distribution of matter, and are thus in many respects analogous to the Newton-Poisson equation $\nabla^2 \Phi = 4\pi G \rho$ which relates the gravitational potential Φ to the matter density ρ, where G is the gravitational constant. But is should be stressed that Einstein's equations amount to rather more than a mere 'relativistic upgrade' of the Newtonian equation—this will become apparent as details of the theory become understood.

The essence of Einstein's theory can be understood as follows. According to the classical theory of continuum mechanics, the equations of motion and the conservation laws for energy, momentum, and angular momentum are embodied in the requirement that a special tensor T^{ab} called the *stress tensor* should be 'conserved'—conserved in the sense that its divergence $\nabla_a T^{ab}$ vanish. Now to grasp this requires something of a leap in both faith and imagination, since the idea applies to a variety of physical theories, and when these theories are cast in their relativistic form the specification of T^{ab} is not always an obviously well-posed problem with a unique answer. Nevertheless for practical purposes one can account for the matter content of space-time by the specification of a symmetric tensor T^{ab} with vanishing divergence. For example, in the case of an ideal fluid we have

$$T^{ab} = (\rho + p)u^a u^b - p g^{ab}$$

where ρ is the energy density, p is the pressure, and u^a is the four-velocity field. The vanishing of the divergence of T^{ab} leads to Euler's equations of motion for the fluid, and to the conservation equations for the energy of the fluid.

Now it turns out that from Γ^a_{bc} one can build up a special tensor G^{ab} that automatically has vanishing divergence. This tensor is called the *Einstein tensor*, and is given explicitly by a simple formula involving terms linear in the first derivatives of Γ^a_{bc} and terms quadratic in Γ^a_{bc} itself. Einstein was led in his investigations to propose that G^{ab}, which is built up geometrically from g_{ab}, must be *proportional* to the stress tensor, and that the factor of the proportionality should be determined by the gravitational constant:

$$G^{ab} = 8\pi G T^{ab}.$$

In this way T^{ab} acts as the 'source' of the gravitational field (as does ρ in the Newton-Poisson equation), whereas the metric g_{ab} thereby determined acts itself on the matter distribution through the requirement that the divergence $\nabla_a T^{ab}$ vanishes where ∇_a is the Levi-Civita connection determined by g_{ab}. And at the same time the matter distribution T^{ab} can depend algebraically on properties of g_{ab}, as seen for example in the case of the fluid stress tensor illustrated above.

Thus Einstein's equations are riddled with non-linearities. This has a number of consequences—not least of which are the difficulties encountered in the construction of exact solutions. And Newton's remarks on the complexities of the many-body problem apply in general relativity even to the two-body problem—since a 'third body' does in effect appear in the form of *gravitational radiation*! Nevertheless many exact solutions are known, and much is known now even about situations where it has not been possible to arrive at a complete description. It is worth bearing in mind that although Einstein's theory remains unchanged in its basic content since its origination in 1915, nevertheless a good deal of work has gone on in the meanwhile, and much is understood now that previously lay shrouded in obscurity, or was simply unknown on account of the lack of appropriate mathematical tools.

But more cannot be said without some systematic development of these tools—a task to which we now turn.

2   Vectors and tensors in flat three-space:
old wine in a new bottle

'I have made a great discovery in mathematics; I have suppressed the summation sign every time that the summation must be made over an index that occurs twice ...'

—Albert Einstein (remark made to a friend)

2.1   Cartesian tensors: an invitation to indices

LOCAL DIFFERENTIAL GEOMETRY consists in the first instance of an amplification and refinement of tensorial methods. In particular, the use of an *index notation* is the key to a great conceptual and geometrical simplification. We begin therefore with a transcription of elementary vector algebra in three dimensions. The ideas will be familiar but the notation new. It will be seen how the index notation gives one insight into the character of relations that otherwise might seem obscure, and at the same time provides a powerful computational tool.

The standard Cartesian coordinates of 3-dimensional space with respect to a fixed origin will be denoted x_i ($i = 1, 2, 3$) and we shall write $\mathbf{A} = A_i$ to indicate that the components of a vector $\mathbf{A}$ with respect to this coordinate system are A_i. The magnitude of $\mathbf{A}$ is given by $\mathbf{A} \cdot \mathbf{A} = A_i A_i$. Here we use the *Einstein summation convention*, whereby in a given term of an expression if an index appears twice an automatic summation is performed: no index may appear more than twice in a given term, and any 'free' (i.e. non-repeated) index is understood to run over the whole range. Thus $A_i A_i$ is an abbreviation for $\sum_i A_i A_i$, and the scalar product between two vectors $\mathbf{A}$ and $\mathbf{B}$ is given by $\mathbf{A} \cdot \mathbf{B} = A_i B_i$.

Multiple index quantities often arise out of problems in geometry and physics. The most basic of these is the *Kronecker delta* δ_{ij} defined by $\delta_{ij} = 1$ if $i = j$ and $\delta_{ij} = 0$ if $i \neq j$. It is essentially the identity matrix, and as a consequence can readily be seen to satisfy $\delta_{ij} = \delta_{ji}$, $\delta_{ij}\delta_{jk} = \delta_{ik}$, $\delta_{ii} = 3$, and $\delta_{ij}A_j = A_i$ for any vector A_i.

Another important multiple index quantity is the permutation tensor or *epsilon tensor*, defined by:

$$\varepsilon_{ijk} = \begin{cases} 1 & \text{if } ijk \text{ is an even permutation of 123} \\ \text{-1} & \text{if } ijk \text{ is an odd permutation of 123} \\ 0 & \text{otherwise, i.e. if } ijk \text{ are not all distinct.} \end{cases}$$

One readily verifies that $\varepsilon_{ijk} = \varepsilon_{jki} = \varepsilon_{kij}$ and that $\varepsilon_{ijk} = -\varepsilon_{jik}$, and $\varepsilon_{iij} = 0$.

Most of the basic identities of vector algebra and vector calculus arise as a consequence of a special relation that holds between δ_{ij} and ε_{ijk}, called the *contracted epsilon identity*:

$$\varepsilon_{iab}\varepsilon_{ipq} = \delta_{ap}\delta_{bq} - \delta_{aq}\delta_{bp}. \tag{2.1.1}$$

Thus when two epsilon tensors are 'contracted' together over their first indices the result can be decomposed into a sum of expressions involving the Kronecker delta. The result is sufficiently basic that it is worth memorizing. Examples of its utility follow forthwith.

The vector product or wedge product $\mathbf{C} = \mathbf{A} \wedge \mathbf{B}$ of two vectors can be expressed by use of epsilon as follows:

$$C_i = \varepsilon_{ijk}A_jB_k. \tag{2.1.2}$$

The scalar triple product of three vectors is

$$\varepsilon_{ijk}P_iQ_jR_k = \mathbf{P} \cdot (\mathbf{Q} \wedge \mathbf{R}) = [\mathbf{P}, \mathbf{Q}, \mathbf{R}]. \tag{2.1.3}$$

Note how the cyclic property of the scalar triple product

$$\mathbf{P} \cdot (\mathbf{Q} \wedge \mathbf{R}) = \mathbf{Q} \cdot (\mathbf{R} \wedge \mathbf{P}) = \mathbf{R} \cdot (\mathbf{P} \wedge \mathbf{Q})$$

follows at once from the expression $\varepsilon_{ijk}P_iQ_jR_k$ by virtue of the identity

$$\varepsilon_{ijk} = \varepsilon_{jki} = \varepsilon_{kij}.$$

In the case of the repeated vector product $\mathbf{A} \wedge (\mathbf{B} \wedge \mathbf{C})$ we derive the following familiar identity:

$$
\begin{aligned}
\mathbf{A} \wedge (\mathbf{B} \wedge \mathbf{C}) &= \varepsilon_{ijk}A_j(\varepsilon_{kpq}B_pC_q) \\
&= \varepsilon_{kij}\varepsilon_{kpq}A_jB_pC_q \\
&= (\delta_{ip}\delta_{jq} - \delta_{iq}\delta_{jp})A_jB_pC_q \\
&= B_iA_qC_q - C_iA_pB_p \\
&= \mathbf{B}(\mathbf{A} \cdot \mathbf{C}) - \mathbf{C}(\mathbf{A} \cdot \mathbf{B}). \tag{2.1.4}
\end{aligned}
$$

Note how simply this identity follows from the contracted epsilon identity. In fact, the argument is reversible, and the known validity of (2.1.4) establishes a proof of (2.1.1). Alternatively, (2.1.1) may be established directly by evaluation component by component.

Further identities may be readily constructed by use of the contracted epsilon identity. For example:

$$
\begin{aligned}
(\mathbf{A} \wedge \mathbf{B}) \cdot (\mathbf{C} \wedge \mathbf{D}) &= (\varepsilon_{ijk}A_jB_k)(\varepsilon_{ipq}C_pD_q) \\
&= \varepsilon_{ijk}\varepsilon_{ipq}A_jB_kC_pD_q \\
&= (\delta_{jp}\delta_{kq} - \delta_{jq}\delta_{kp})A_jB_kC_pD_q \\
&= A_pC_pB_qD_q - A_qD_qB_pC_p \\
&= (\mathbf{A} \cdot \mathbf{C})(\mathbf{B} \cdot \mathbf{D}) - (\mathbf{A} \cdot \mathbf{D})(\mathbf{B} \cdot \mathbf{C}). \tag{2.1.5}
\end{aligned}
$$

There is another identity involving epsilon which allows one to express an outer product of two epsilons (i.e. a product involving no contractions) in terms of Kronecker deltas:

$$\varepsilon_{ijk}\varepsilon_{pqr} = \delta_{ip}\delta_{jq}\delta_{kr} + \delta_{iq}\delta_{jr}\delta_{kp} + \delta_{ir}\delta_{jp}\delta_{kq}$$
$$- \delta_{iq}\delta_{jp}\delta_{kr} - \delta_{ir}\delta_{jq}\delta_{kp} - \delta_{ip}\delta_{jr}\delta_{kq}. \tag{2.1.6}$$

Identity (2.1.1) follows as a special case of (2.1.6) via contraction on the left side of the equation over the first index on each epsilon. As an example of the use of (2.1.6) observe that if we transvect each side with $A_i B_j C_k A_p B_q C_r$ we obtain

$$[\mathbf{A},\mathbf{B},\mathbf{C}]^2 = \mathbf{A}^2\mathbf{B}^2\mathbf{C}^2 + 2(\mathbf{A}\cdot\mathbf{B})(\mathbf{B}\cdot\mathbf{C})(\mathbf{C}\cdot\mathbf{A})$$
$$- (\mathbf{A}\cdot\mathbf{B})^2\mathbf{C}^2 - (\mathbf{A}\cdot\mathbf{C})^2\mathbf{B}^2 - (\mathbf{B}\cdot\mathbf{C})^2\mathbf{A}^2. \tag{2.1.7}$$

If the vectors $\mathbf{A}$, $\mathbf{B}$, and $\mathbf{C}$ join at a vertex of a parallelepiped, then $[\mathbf{A},\mathbf{B},\mathbf{C}]$ is the volume of this solid, and as a consequence if A, B, C are the lengths of $\mathbf{A}$, $\mathbf{B}$, $\mathbf{C}$, and α, β, γ the angles between them, then formula (2.1.7) implies that the volume of the parallelepiped is:

$$ABC[1 + 2\cos\alpha\cos\beta\cos\gamma - \cos^2\alpha - \cos^2\beta - \cos^2\gamma]^{1/2} \tag{2.1.8}$$

by use of the relations $A_iB_i = AB\cos\alpha$, $B_iC_i = BC\cos\beta$, and $C_iA_i = CA\cos\gamma$. Note, incidentally, that (2.1.7) may be also expressed in the form:

$$[\mathbf{A},\mathbf{B},\mathbf{C}]^2 = \begin{vmatrix} \mathbf{A}\cdot\mathbf{A} & \mathbf{A}\cdot\mathbf{B} & \mathbf{A}\cdot\mathbf{C} \\ \mathbf{B}\cdot\mathbf{A} & \mathbf{B}\cdot\mathbf{B} & \mathbf{B}\cdot\mathbf{C} \\ \mathbf{C}\cdot\mathbf{A} & \mathbf{C}\cdot\mathbf{B} & \mathbf{C}\cdot\mathbf{C} \end{vmatrix} \tag{2.1.9}$$

which may be directly verified by calculation with respect to a suitable basis. We may regard (2.1.9) therefore as establishing a proof of the identity (2.1.6).

2.2   Matrix algebra made easy

As a further illustration of the utility of the index notation let us consider the algebra of three-by-three matrices. Again, much of the material will be familiar from other contexts, but what should be evident and taken on board is the way index methods provide a neat and compact way of deducing many of the relevant relations. (Most of the formulae can be straightforwardly generalized to the n by n case.)

Let U_{ij} be a three-by-three matrix, where the first index labels rows and the second index labels columns, and let Δ denote its determinant. There is an elegant formula for Δ in terms of U_{ij} and ε_{ijk} given as follows:

$$\Delta = \frac{1}{6}\varepsilon_{ijk}U_{ip}U_{jq}U_{kr}\varepsilon_{pqr}. \tag{2.2.1}$$

It is helpful to have at our disposal two further identities, namely:

$$\Delta\varepsilon_{ijk} = \varepsilon_{pqr}U_{ip}U_{jq}U_{kr}, \tag{2.2.2}$$
$$\Delta\varepsilon_{pqr} = \varepsilon_{ijk}U_{ip}U_{jq}U_{kr}. \tag{2.2.3}$$

From these expressions most of the familiar properties of determinants can be deduced. For example, if we take the $ijk = 123$ component of (2.2.2), and write $A_p = U_{1p}$, $B_q = U_{2q}$, $C_r = U_{3r}$, we get:

$$\varepsilon_{pqr} A_p B_q C_r = \begin{vmatrix} A_1 & A_2 & A_3 \\ B_1 & B_2 & B_3 \\ C_1 & C_2 & C_3 \end{vmatrix}, \tag{2.2.4}$$

a well-known formula for the scalar triple product. It follows moreover that Δ vanishes if and only if one of its rows is linearly dependent on the other two, e.g. $C_r = \alpha A_r + \beta B_r$ for some α, β. Formula (2.2.1) is obtained by contraction of (2.2.2) with ε_{ijk}, bearing in mind that $\varepsilon_{ijk}\varepsilon_{ijk} = 6$.

Matrix multiplication is given as follows: if β_{ij} and γ_{jk} are matrices then their matrix product is $\alpha_{ik} = \beta_{ij}\gamma_{jk}$, with use of the summation convention of course. A straightforward exercise verifies this is equivalent to the standard procedure. As a further example of the use of the identities (2.2.2) and (2.2.3) we shall deduce the *multiplicative law of determinants*: viz., if $\alpha_{ik} = \beta_{ij}\gamma_{jk}$ then $\Delta_\alpha = \Delta_\beta \Delta_\gamma$ where $\Delta_\alpha = det(\alpha_{ij})$. We have:

$$\begin{aligned} \varepsilon_{ijk}\Delta_\alpha &= \alpha_{ip}\alpha_{jq}\alpha_{kr}\varepsilon_{pqr} \\ &= (\beta_{is}\gamma_{sp})(\beta_{jt}\gamma_{tq})(\beta_{ku}\gamma_{ur})\varepsilon_{pqr} \\ &= \beta_{is}\beta_{jt}\beta_{ku}(\gamma_{sp}\gamma_{tq}\gamma_{ur}\varepsilon_{pqr}) \\ &= \beta_{is}\beta_{jt}\beta_{ku}\varepsilon_{stu}\Delta_\gamma \\ &= \varepsilon_{ijk}\Delta_\beta\Delta_\gamma. \end{aligned} \tag{2.2.5}$$

The *inverse* of a matrix, when it exists, has a straightforward representation by use of ε_{ijk}. As before let $\Delta = det(U_{ij})$, and assume $\Delta \neq 0$. If we define

$$V_{jk} = (2\Delta)^{-1}\varepsilon_{jpq}\varepsilon_{kab}U_{ap}U_{bq} \tag{2.2.6}$$

then we obtain $U_{ij}V_{jk} = \delta_{ik}$ showing thereby that V_{jk} is the inverse of U_{ij}. The proof proceeds as follows:

$$\begin{aligned} U_{ij}V_{jk} &= U_{ij}(2\Delta)^{-1}\varepsilon_{jpq}\varepsilon_{kab}U_{ap}U_{bq} \\ &= (2\Delta)^{-1}\varepsilon_{kab}(U_{ij}U_{ap}U_{bq}\varepsilon_{jpq}) \\ &= (2\Delta)^{-1}\varepsilon_{kab}(\varepsilon_{iab}\Delta) \\ &= \delta_{ik}, \end{aligned} \tag{2.2.7}$$

where in the last step we use the identity $\varepsilon_{iab}\varepsilon_{kab} = 2\delta_{ik}$.

2.3   Vector calculus seen afresh

As a convenient abbreviation we write $\nabla_i \phi = \partial\phi/\partial x^i$ for the vector derivative of ϕ. If ϕ has continuous second derivatives then the commutativity of partial differentiation on ϕ, given by

$$(\nabla_j \nabla_k - \nabla_k \nabla_j)\phi = 0, \tag{2.3.1}$$

can by use of the ε_{ijk} tensor be expressed succinctly in the form

$$\varepsilon_{ijk}\nabla_j\nabla_k\phi = 0. \tag{2.3.2}$$

Indeed, (2.3.2) can be obtained by contraction of (2.3.1) with ε_{ijk}. Conversely, if we contract (2.3.2) with ε_{iab} we get back to (2.3.1) by use of (2.1.1).

The *divergence, gradient*, and *curl* operations are formulated as follows:

$$div\,\mathbf{V} = \nabla\cdot\mathbf{V} = \nabla_i V_i$$

$$grad\,\phi = \nabla\phi = \nabla_i\phi$$

$$curl\,\mathbf{V} = \nabla\wedge\mathbf{V} = \varepsilon_{ijk}\nabla_j V_k \tag{2.3.3}$$

for any sufficiently differentiable field V_i. We shall use the notation $\nabla^2\phi = \nabla\cdot\nabla\phi = \nabla_i\nabla_i\phi$ for the Laplacian operator.

A number of identities follow on directly from these definitions. First we have the two 'trivial' vector calculus identities

$$div(curl\,\mathbf{V}) = \nabla_i(\varepsilon_{ijk}\nabla_j V_k) = 0$$

$$curl(grad\,\phi) = \varepsilon_{ijk}\nabla_j(\nabla_k\phi) = 0 \tag{2.3.4}$$

which follow as a consequence of the commutation relations (2.3.2) for ∇_i. More sophisticated identities then follow as in the case of vector algebra, for example:

$$
\begin{aligned}
\nabla\wedge(\nabla\wedge\mathbf{V}) &= \varepsilon_{ijk}\nabla_j(\varepsilon_{kpq}\nabla_p V_q)\\
&= \varepsilon_{ijk}\varepsilon_{kpq}\nabla_j\nabla_p V_q\\
&= (\delta_{ip}\delta_{jq}-\delta_{iq}\delta_{jp})\nabla_j\nabla_p V_q\\
&= \nabla_i(\nabla_j V_j)-(\nabla_j\nabla_j)V_i\\
&= \nabla(\nabla\cdot\mathbf{V})-\nabla^2\mathbf{V},
\end{aligned}
\tag{2.3.5}
$$

and also

$$
\begin{aligned}
\nabla\cdot(\mathbf{A}\wedge\mathbf{B}) &= \nabla_i(\varepsilon_{ijk}A_j B_k)\\
&= B_k(\nabla_i\varepsilon_{ijk}A_j)+A_j(\nabla_i\varepsilon_{ijk}B_k)\\
&= B_k(\varepsilon_{kij}\nabla_i A_j)-A_j(\varepsilon_{jik}\nabla_i B_k)\\
&= \mathbf{B}\cdot(\nabla\wedge\mathbf{A})-\mathbf{A}\cdot(\nabla\wedge\mathbf{B}).
\end{aligned}
\tag{2.3.6}
$$

Further such identities can be readily generated. For instance, if we write $\omega_i = \varepsilon_{ijk}\nabla_j V_k$ for the curl of the vector field V_k then:

$$V_j\nabla_j V_i = \frac{1}{2}\nabla_i V^2 - \varepsilon_{ijk}V_j\omega_k, \tag{2.3.7}$$

an identity that is very useful in non-relativistic fluid mechanics. The proof of relation (2.3.7) is:

$$
\begin{aligned}
\varepsilon_{ijk}V_j\omega_k &= \varepsilon_{ijk}V_j(\varepsilon_{kpq}\nabla_p V_q)\\
&= \varepsilon_{ijk}\varepsilon_{kpq}V_j\nabla_p V_q\\
&= (\delta_{ip}\delta_{jq}-\delta_{iq}\delta_{jp})V_j\nabla_p V_q\\
&= V_j\nabla_i V_j - V_j\nabla_j V_i\\
&= \frac{1}{2}\nabla_i(V_j V_j) - V_j\nabla_j V_i.
\end{aligned}
\tag{2.3.8}
$$

The basic theorems of vector *integral* calculus may be summarised in a slightly abstract form by the following three formulae:

$$\int_\gamma \nabla_i(Q) dx_i = (Q)_B - (Q)_A \tag{2.3.9}$$

$$\int_S \varepsilon_{ijk} N_i \nabla_j(Q) dS = \int_\Gamma (Q) dx_k \tag{2.3.10}$$

$$\int_V \nabla_i(Q) dV = \int_\Sigma (Q) N_i dS. \tag{2.3.11}$$

Here Q is any smooth field, which may be taken variously to be a scalar field, a vector field, or indeed a multiple-index quantity. Contractions, where appropriate, are allowed between an index on Q and the free index in (2.3.10) or (2.3.11).

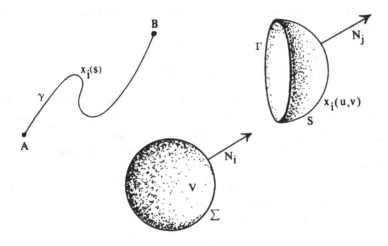

Figure 2.1. Integral theorems of vector calculus.

In formula (2.3.9) the integration is along a smooth curve γ with end points A and B. If the curve is given parametrically by $x_i = x_i(s)$ where s is arc *length* then $dx_i = t_i ds$ where $t_i(s)$ is the *unit tangent vector* along the curve $(t_i t_i = 1)$.

In formula (2.3.10) S is a simple smooth surface bounded by a smooth curve Γ, N_j is the unit normal vector to the surface, and dS the area element. If the surface is represented parametrically according to the scheme $x_i = x_i(u,v)$ then N_j and dS are determined by the relation

$$N_i dS = \varepsilon_{ijk} U_j V_k du dv, \tag{2.3.12}$$

where $U_j = \partial x_j/\partial u$ and $V_j = \partial x_j/\partial v$. Since N_i must be a unit vector we therefore have $dS = \Psi(u,v) du dv$ and $N_i = \Psi^{-1} \varepsilon_{ijk} U_j V_k$ where Ψ is given by $\Psi^2 = \mathbf{U}^2 \mathbf{V}^2 - (\mathbf{U} \cdot \mathbf{V})^2$, after use of (2.1.5).

And in formula (2.3.11) V is a region of space bounded by a closed smooth surface Σ. N_i is the outward-pointing vector field normal to Σ and dV the volume element.

As an elementary application of (2.3.9) we let Q be a gravitational potential Φ. If m is the mass of a point particle moving under the influence of Φ, then $-m\nabla_i\Phi$ is the gravitational force acting on it, and $-\int_\gamma m\nabla_i\Phi dx_i$ is the energy required to move the particle along the path γ from A to B. We see by use of (2.3.9) that the energy expended in moving a particle from A to B in a gravitational field depends only on the difference of the values of the potential at the end points.

If in (2.3.10) we set $Q = A_k$ then we obtain the Stokes theorem, and in (2.3.11) if we set $Q = A_k$ we obtain the Gauss theorem. With other choices for Q we then obtain a number of well-known variants on these theorems.

There are two further results in vector integral calculus which are quite useful that arise (non-trivially) as converses of the trivial identities (2.3.4): if a vector field ω_i satisfies $\nabla_i\omega_i = 0$ then locally there exists a vector field V_k such that $\omega_i = \varepsilon_{ijk}\nabla_j V_k$, and if a vector field V_k satisfies $\varepsilon_{ijk}\nabla_j V_k = 0$ then there exists locally a scalar ϕ such that $V_k = \nabla_k\phi$.

2.4   Elementary electromagnetism

It is of interest to formulate in the index notation some simple examples of physical systems, with a view particularly to the corresponding formulations in four dimensions in later chapters. First we consider electromagnetism. We assume here a basic familiarity on the part of the reader with electromagnetic theory—but otherwise that which follows may provide a useful introduction.

Let E_i be the electric field, B_i the magnetic field, ρ the electric charge density, and J_i the electric current density. Maxwell's equations, the fundamental differential equations of electromagnetic theory, are then given as follows:

$$\nabla_i B_i = 0 \tag{2.4.1}$$

$$\nabla_i E_i = 4\pi\rho \tag{2.4.2}$$

$$\varepsilon_{ijk}\nabla_j E_k = -\dot{B}_i \tag{2.4.3}$$

$$\varepsilon_{ijk}\nabla_j B_k = \dot{E}_i + 4\pi J_i \tag{2.4.4}$$

where the dot denotes the time derivative $\partial/\partial t$. As a consequence of these relations we are assured of the existence of a scalar potential ϕ and a vector potential A_i such that

$$B_i = \varepsilon_{ijk}\nabla_j A_k \tag{2.4.5}$$

$$E_i = -\dot{A}_i - \nabla_i\phi \tag{2.4.6}$$

where A_i and ϕ are determined up to a 'gauge transformation' $A_i \rightarrow A_i + \nabla_i\psi$ and $\phi \rightarrow \phi - \dot{\psi}$ where ψ is an arbitrary function.

Indeed the existence of A_k such that (2.4.5) holds follows directly from (2.4.1), and if we substitute this expression for B_i into (2.4.3) we obtain $\varepsilon_{ijk}\nabla_j C_k = 0$ where $C_k = E_k + \dot{A}_k$, from which it follows at once that $C_k = \nabla_k \phi$ for some ϕ.

If we differentiate (2.4.2) with respect to time and take the divergence of (2.4.4) then on comparison of the results we obtain

$$\nabla_i J_i + \dot{\rho} = 0, \tag{2.4.7}$$

which is the equation of conservation for electric charge. This interpretation of (2.4.7) follows from the fact that the net electric charge Q in a region V is given by

$$Q = \int_V \rho dV \tag{2.4.8}$$

from which we deduce that

$$\dot{Q} = \int_V \dot{\rho} dV = -\int_V \nabla_i J_i dV = -\int_\Sigma J_i N_i dS \tag{2.4.9}$$

by use of (2.4.7) and the Gauss theorem (2.3.11), showing that the rate of increase of Q is equal to the net flux of charge across Σ into V, where Σ is the boundary of V.

Note therefore, incidentally, that any equation of the form (2.4.7) may be interpreted as a conservation law, provided ρ and J_i are appropriately identified as the density and flux of the conserved quantity.

Some further relations of interest follow on in the special case of a *source-free* or vacuum electromagnetic field, given by $\rho = 0$, $J_i = 0$. In particular we find that E_i and B_i each satisfy the *wave equation* $\ddot{E}_i = \nabla^2 E_i$ and $\ddot{B}_i = \nabla^2 B_i$. The first of these relations is obtained, for example, by comparison of the curl of (2.4.3) with the time derivative of (2.4.4) followed by use of the identity (2.3.5).

Another important relation can be obtained from the vacuum equations by contraction of (2.4.3) with B_i and (2.4.4) with E_i. The resulting equations, on combination, and by use of (2.3.6), yield:

$$\nabla_i P_i + \dot{\mu} = 0 \tag{2.4.10}$$

where $8\pi\mu = E^2 + B^2$ and $4\pi P_i = \varepsilon_{ijk}E_j B_k$. The scalar μ is the *density of electromagnetic energy*, and the vector P_i is the *flux density* (current) of the electromagnetic energy. Equation (2.4.10) expresses the fact that in a vacuum the energy of the electromagnetic field is conserved.

Now suppose a particle of mass m and charge q moves under the influence of an electromagnetic field E_i, B_i. The classical law of motion governing it is

$$m\dot{V}_i = q(E_i + \varepsilon_{ijk}V_j B_k), \tag{2.4.11}$$

which is called the Lorentz equation. Here $V_i = dx^i/dt$ denotes the tangent to the trajectory $x^i(t)$ at time t.

Throughout the discussion we have slipped in (you may not have noticed) an important 'relativistic' convention—namely, the speed of light $c \approx 2.9979 \times 10^{10}$ *cm* s^{-1}

has been set to unity. This is very natural for the study of electromagnetism, even in a non-relativistic setting. Units of length and time can be freely interchanged by use of the conversion factor 2.9979×10^{10} *cm* ≈ 1 *s.*

The basic unit of charge is the so-called *electrostatic unit*. This is given by

$$1 \; esu = 1 \; g^{1/2} \; cm^{3/2} \; s^{-1}. \tag{2.4.12}$$

The fractional 'dimensions' may seem odd, but they are essential. The electrostatic unit is defined so that two particles each charged with one *esu* will, when placed 1 *cm* apart, repel each other with a force of one *dyne* (1 *g cm* s^{-2}).

The charge on the electron is approximately -4.80325×10^{-10} *esu.* It is a straight-forward exercise to verify that if e is the electron charge, then $e^2 = \alpha \hbar c$, where $\hbar = h/2\pi = 1.05459 \times 10^{-27}$ *g cm² s⁻¹* is Planck's constant (divided by 2π), and $\alpha = 7.29735 \times 10^{-3}$ is the so-called *fine structure constant* ($\alpha^{-1} = 137.0360$).

The *esu* is the preferred unit of electromagnetic charge for 'microscopic' applications, and is used very nearly universally in atomic physics, nuclear physics, particle physics, plasma physics, and astrophysics, as well as relativity theory. For 'macroscopic' applications, particularly engineering and electronics, we use the *coulomb* as a unit of charge, given approximately by

$$1 \; \text{coulomb} = 2.9979 \times 10^9 \; esu. \tag{2.4.13}$$

In units such that the speed of light is one we have

$$1 \; \text{coulomb} = \frac{1}{10} g^{1/2} cm^{1/2}. \tag{2.4.14}$$

The factor of $\frac{1}{10}$ is a bit awkward, but that's not our fault.

2.5   Non-relativistic fluid motion

The Newtonian mechanics of point particles can be generalized simply and naturally in two ways: to rigid body mechanics, and to fluid mechanics. The former is strangely refractory when it comes to a relativistic formulation, whereas the latter can be formulated very elegantly in relativistic terms, and forms the basis of a number of important relativistic cosmological and astrophysical investigations.

Here we shall review some elementary aspects of non-relativistic fluid motion, casting the theory into the index notation.

An *ideal* or *perfect* fluid is described by its density ρ, its pressure p, and its velocity field V_i, all taken to be functions both of space and time. The equations of motion of a fluid, Euler's equations, are

$$\rho[\dot{V_i} + V_j \nabla_j V_i] = -\nabla_i p. \tag{2.5.1}$$

This equation is supplemented by the further relation

$$\nabla_i (\rho V_i) + \dot{\rho} = 0 \tag{2.5.2}$$

which expresses the conservation of mass (cf. equation 2.4.7). These differential equations must be solved subject to appropriate initial conditions and boundary conditions, and physical equations relating ρ and p that depend on specific characteristics of the particular fluid under study: thermodynamic considerations may also have to be brought into play.

The *vorticity* ω_i is defined by $\omega_i = \varepsilon_{ijk}\nabla_j V_k$. By use of identity (2.3.7) Euler's equation may be cast into the alternative form

$$\rho \dot{V}_i = \rho \varepsilon_{ijk} V_j \omega_k - \frac{1}{2}\rho \nabla_i V^2 - \nabla_i p \qquad (2.5.3)$$

from which it is possible in special cases to derive a number of interesting conclusions.

A *steady* flow is one for which ρ, p, and V_i are all time independent. For a steady flow with $\omega_i = 0$ (zero vorticity) we obtain

$$\frac{1}{2}\rho \nabla_i V^2 + \nabla_i p = 0 \qquad (2.5.4)$$

from which it follows that ρ can be expressed as a function of p, and hence therefore that there exists a function h such that $\nabla_i(\rho^{-1}h) = \rho^{-1}\nabla_i p$. Equation (2.5.4) can then be integrated to yield

$$\frac{1}{2}\rho V^2 + h = E \qquad (2.5.5)$$

where E is constant. If we further specialize to the case of an *incompressible fluid* (ρ constant) then $h = p$ and equation (2.5.5) expresses *Bernoulli's theorem*. Since $\omega_i = 0$ we have $V_i = \nabla_i \Phi$ where Φ is the *velocity potential*, and equation (2.5.2) reduces to $\nabla^2 \Phi = 0$. Thus *steady irrotational incompressible flows are characterized by solutions of Laplace's equation.*

It should be evident that in order for an energy integral of the form (2.5.5) to be obtained it is not strictly necessary that ω_i should vanish, merely that $\varepsilon_{ijk} V_j \omega_k$ should, which is a necessary and sufficient condition that the fluid is 'foliated' locally by a family of 2-surfaces to which the streamlines of the flow are orthogonal; i.e. $V_i = \alpha \nabla_i \beta$ for suitable scalars α and β, the 'leaves' of the foliation being surfaces of constant β.

In the case of a *time-dependent incompressible flow* an important formula can be obtained characterizing the rate of change of vorticity. If we assume ρ to be constant and take the curl of (2.5.3) it follows that

$$\begin{aligned}
\dot{\omega}_i &= \varepsilon_{ijk}\nabla_j(\varepsilon_{kpq}V_p\omega_q) \\
&= \varepsilon_{kij}\varepsilon_{kpq}\nabla_j(V_p\omega_q) \\
&= (\delta_{ip}\delta_{jq} - \delta_{iq}\delta_{jp})(\omega_q\nabla_j V_p + V_p\nabla_j\omega_q) \\
&= \omega_j\nabla_j V_i - \omega_i\nabla_j V_j + V_i\nabla_j\omega_j - V_j\nabla_j\omega_i.
\end{aligned} \qquad (2.5.6)$$

Now $\nabla_j\omega_j$ vanishes automatically by (2.3.4), and $\nabla_j V_j$ vanishes by virtue of the conservation equation (2.5.2) since ρ is assumed constant; thus we are left with

$$\dot{\omega}_i + \zeta_i = 0 \qquad (2.5.7)$$

where the *commutator* ζ_i is defined by

$$\zeta_i = V_j \nabla_j \omega_i - \omega_j \nabla_j V_i. \tag{2.5.8}$$

Equation (2.5.7) expresses the content of *Helmholtz's theorem* for ideal fluids, the relativistic analogue of which we shall meet later (exercise 8.3). Note again how the index notation renders these calculations elegant and tractable—expressions that would be ungainly otherwise can by tensor methods be written concisely and straightforwardly. This is of considerable importance in relativistic theory, where expressions of some complexity often arise.

Equation (2.5.2) is the conservation law for the *mass* of the fluid, so it is natural to ask for a corresponding conservation law for the *momentum* of the fluid; this is in fact expressed by equation (2.5.1), although this may not be evident from the form in which the equation is written.

To cast (2.5.1) into a form where it is manifestly a momentum conservation law it is helpful to introduce another tensor T_{ij} called the *stress tensor* of the fluid, defined by

$$T_{ij} = \rho V_i V_j + p \delta_{ij}. \tag{2.5.9}$$

Equation (2.5.1) may then be expressed in the form

$$\partial_t (\rho V_i) + \nabla_j T_{ij} = 0 \tag{2.5.10}$$

which for each value of i gives a flux law for the corresponding momentum component. To see that (2.5.10) is indeed equivalent to (2.5.1) we proceed according to the following calculations:

$$\begin{aligned}
\partial_t (\rho V_i) + \nabla_j T_{ij} &= \dot{\rho} V_i + \rho \dot{V_i} + \nabla_j (\rho V_i V_j + p \delta_{ij}) \\
&= -\nabla_j (\rho V_j) V_i + \rho \dot{V_i} + \nabla_j (\rho V_j) V_i \\
&\quad + \rho V_j \nabla_j V_i + \nabla_i p \\
&= \rho \dot{V_i} + \rho V_j \nabla_j V_i + \nabla_i p
\end{aligned} \tag{2.5.11}$$

where use is made of the continuity equation (2.5.2) in going to the second line of the calculation.

2.6 Self-gravitating fluids

We may enquire how (2.5.1) is to be modified when the fluid moves under the influence of a gravitational field. If Φ is the gravitational potential then a force term is included in (2.5.1) so that it reads

$$\rho [\dot{V_i} + V_j \nabla_j V_i] = -\nabla_i p - \rho \nabla_i \Phi \tag{2.6.1}$$

or equivalently

$$\partial_t (\rho V_i) + \nabla_j T_{ij} = -\rho \nabla_i \Phi \tag{2.6.2}$$

with T_{ij} given as in (2.5.9). A *self gravitating* fluid (such as might be used to model a star) is one for which the mass density of the fluid is itself the source of the gravitational field. In this case we supplement (2.5.2) and (2.6.2) with the Newton-Poisson equation for the gravitational potential, given by

$$\nabla^2 \Phi = 4\pi G \rho \tag{2.6.3}$$

where G is the gravitational constant.

It is interesting in the case of a self-gravitating fluid to ask whether T_{ij} can be augmented by a term τ_{ij} representing 'gravitational stress' in such a way that (2.6.2) can be written in the form

$$\partial_t(\rho V_i) + \nabla_j(T_{ij} + \tau_{ij}) = 0 \tag{2.6.4}$$

thereby expressing conservation of the momentum of the fluid when it moves under the influence of its own gravitational field. The following expression can be deduced:

$$4\pi G \tau_{ij} = \nabla_i \Phi \nabla_j \Phi - \frac{1}{2}\delta_{ij}\nabla_k \Phi \nabla_k \Phi \tag{2.6.5}$$

which upon taking its divergence yields:

$$
\begin{aligned}
4\pi G \nabla_j \tau_{ij} &= \nabla_j(\nabla_i \Phi \nabla_j \Phi) - \frac{1}{2}\nabla_i(\nabla_j \Phi \nabla_j \Phi) \\
&= \nabla_i \Phi(\nabla_j \nabla_j \Phi) + \nabla_i(\nabla_j \Phi)\nabla_j \Phi - (\nabla_i \nabla_j \Phi)\nabla_j \Phi \\
&= \nabla_i \Phi(4\pi G \rho)
\end{aligned}
\tag{2.6.6}
$$

as desired, where in going to the last line we use the Newton-Poisson equation.

We see that the use of the stress tensor, a symmetric three-by-three array built up from the fluid density, the pressure, and the velocity, is very helpful in deducing certain properties of the fluid—properties that cannot be expressed very conveniently otherwise. Here we have quite a good example of the application of tensors as indexed quantities playing a simplifying role in the description of a physical system. The notion of a 'stress tensor' is of considerable importance in its own right, and comes into play in a major way in the formulation of Einstein's equations for the gravitational field.

Exercises for chapter 2

[2.1] Establish the following identity by index methods:
$$(\mathbf{A} \wedge \mathbf{B}) \wedge (\mathbf{P} \wedge \mathbf{Q}) = -\mathbf{A}[\mathbf{B}, \mathbf{P}, \mathbf{Q}] + \mathbf{B}[\mathbf{A}, \mathbf{P}, \mathbf{Q}] = \mathbf{P}[\mathbf{Q}, \mathbf{A}, \mathbf{B}] - \mathbf{Q}[\mathbf{P}, \mathbf{A}, \mathbf{B}]$$

[2.2] Establish the following identities:
    (i) $div(grad\,\phi \wedge grad\,\psi) = 0$

(ii) $curl(\mathbf{A} \wedge \mathbf{B}) = \mathbf{A}\,div\,\mathbf{B} - \mathbf{B}\,div\,\mathbf{A} + \mathbf{B}\cdot\nabla\mathbf{A} - \mathbf{A}\cdot\nabla\mathbf{B}$

[2.3] Solve for X_i:
$$kX_i + \varepsilon_{ijk}X_jP_k = Q_i$$

[2.4] Solve for X_i and Y_i:
$$aX_i + \varepsilon_{ijk}Y_jP_k = A_i$$
$$bY_i + \varepsilon_{ijk}X_jP_k = B_i$$

[2.5] Solve for X_i:
$$\varepsilon_{ijk}X_jA_k = B_i \quad X_iC_i = k \quad (A_iB_i = 0)$$

[2.6] Solve for X_i and Y_i:
$$aX_i + bY_i = P_i$$
$$\varepsilon_{ijk}X_jY_k = Q_i, \quad (P_iQ_i = 0)$$

[2.7] Solve: $\varepsilon_{ijk}X_jP_k + X_jQ_jR_i = S_i$

[2.8] Solve: $kX_i + \varepsilon_{ijk}X_jP_k + X_jQ_jR_i = S_i$

[2.9] In n-dimensional space the epsilon tensor $\varepsilon_{ab...pq}$ has n indices and is anti-symmetric under the interchange of any two adjacent indices with $\varepsilon_{12...n} = 1$. Deduce appropriate formulae for the determinant and the inverse of an n by n matrix.

[2.10] The *transpose* of a matrix A_{ij} is the matrix $\tilde{A}_{ij} = A_{ji}$. A matrix is called *symmetric* if $A_{ij} = \tilde{A}_{ij}$, and *anti-symmetric* if $A_{ij} = -\tilde{A}_{ij}$. A matrix A_{ij} is called *orthogonal* if $A_{ij}\tilde{A}_{jk} = \delta_{ik}$. Show that if A_{ij} and B_{ij} are both orthogonal matrices then their product $C_{ik} = A_{ij}B_{jk}$ is also orthogonal. Suppose P_{ij} is anti-symmetric, and that $\delta_{ij}+P_{ij}$ has an inverse Q_{ij}. Show that $A_{ik} = (\delta_{ij}-P_{ij})Q_{jk}$ is orthogonal.

[2.11] A matrix is called *Hermitian* if $\bar{A}_{ij} = \tilde{A}_{ij}$, where the bar denotes complex conjugation, and *anti-Hermitian* if $\bar{A}_{ij} = -\tilde{A}_{ij}$. A matrix is called *unitary* if $U_{ij}\tilde{\bar{U}}_{jk} = \delta_{ik}$. Show that the product of two unitary matrices is unitary. Suppose H_{ij} is anti-Hermitian, and that $\delta_{ij} + H_{ij}$ has an inverse K_{ij}. Show that $U_{ik} = (\delta_{ij} - H_{ij})K_{jk}$ is unitary.

[2.12] Verify that if E_i and B_i satisfy Maxwell's equations *in vacuo* then they each satisfy the wave equation. Give necessary and sufficient conditions on ρ and J_i for E_i and B_i to satisfy the wave equation.

[2.13] Suppose that E_i and B_i can each be developed into a power series in time:
$$E_i = \sum t^n E_i^n, \quad B_i = \sum t^n B_i^n$$
where the index n (not a tensor index) runs from zero to infinity; E_i^n and B_i^n depend only on the spatial co-ordinates. Under the assumption that the relevant series converge show that if E_i^0 and B_i^0 are specified, subject to $\nabla^i E_i^0 = 0$ and $\nabla_i B_i^0 = 0$, then Maxwell's equations *in vacuo* determine E_i and B_i uniquely.

[2.14] Generalize the result of exercise [2.13] to the case when charge and current are present.

[2.15] Find the general solution of the Lorentz equation (2.4.11) for constant E_i and B_i, given the initial position and velocity of the particle.

[2.16] The *electromagnetic stress tensor* T_{ij} is defined according to the formula:
$$4\pi T_{ij} = -E_i E_j - B_i B_j + \frac{1}{2}\delta_{ij}(E_k E_k + B_k B_k).$$
Show that in a vacuum it follows as a consequence of Maxwell's equations that
$$\dot{P}_i + \nabla_j T_{ij} = 0$$
where P_i is the electromagnetic energy flux vector defined by
$$4\pi P_i = \varepsilon_{ijk} E_j B_k.$$

[2.17] Show that equation (2.5.7), which was derived under the assumption of an *incompressible* flow, holds in fact for any inviscid fluid for which ρ can be expressed as a function of p.

[2.18] If *viscous* effects are included in the analysis of fluid dynamics then Euler's equations are modified to read
$$\rho[\dot{V}_i + V_j \nabla_j V_i] = -\nabla_i p + \eta \nabla^2 V_i + (\zeta + \frac{1}{3}\eta)\nabla_i(\nabla_j V_j)$$
where η is the *dynamic viscosity*, and ζ the *bulk viscosity*; here both are assumed constant. In the case of an incompressible fluid (ρ constant) we have $\nabla_j V_j = 0$ by the continuity equation and we obtain the *Navier-Stokes equation*. Verify in this case that the vorticity ω_i satisfies a Helmholtz relation in the form of the following diffusion equation:
$$\dot{\omega}_i - \nu\nabla^2\omega_i = \zeta_i$$
where $\nu = \eta/\rho$ is the so-called *kinematic viscosity*.

[2.19] Show that the general equation for viscous fluid flow can be represented in the form:
$$\partial_t(\rho V_i) + \nabla_j T_{ij} = 0$$
where
$$T_{ij} = \rho V_i V_j + (p - \zeta\nabla_k V_k)\delta_{ij} - 2\eta[\nabla_{(i}V_{j)} - \frac{1}{3}\delta_{ij}\nabla_k V_k].$$

3 Aspects of special relativistic geometry

The circumstance that there is no objective rational division of the four-dimensional continuum into a three-dimensional space and a one-dimensional time continuum indicates that the laws of nature will assume a form which is logically most satisfactory when expressed as laws in the four-dimensional space-time continuum. Upon this depends the great advance in method which the theory of relativity owes to Minkowksi.

—Albert Einstein (**The Meaning of Relativity**)

3.1   Light-cone geometry: the key to special relativity

WE HAVE SEEN how an index notation is strikingly helpful in the development of physical formulae for flat three-dimensional space. We found it convenient to work with a fixed Cartesian coordinate system, expressing the components of vectors and tensors with respect to that system. We know, nevertheless, as a matter of principle, that the general conclusions we draw are independent of the particular coordinatization chosen for the underlying space.

We now propose to formulate special relativity in essentially the same spirit. We shall regard space-time as a flat four-dimensional continuum with coordinates x^a ($a = 0, 1, 2, 3$). The points of space-time are called 'events', and we are interested in the relations of events to one another. Our purpose here is two-fold: first, to review some aspects of special relativity pertinent to that which follows later; and second, to develop further a number of index-calculus tools which are very useful in general relativity as well as special relativity.

We assume the reader to be already familiar with at least the rudiments of special relativity, so no particular attempt will be made here to develop the theory from first principles. We also presume the reader to have had at least some tentative exposure to the formulation of special relativistic ideas in the language of vectors and tensors in four dimensions, so that although the details in the following discussion may be new, nevertheless the framework in general outline will already be to some extent familiar.

Let us proceed now to summarise the elementary details of vector and tensor algebra in four dimensions. Relative to a fixed set of coordinates x^a in space-time ($a = 0, 1, 2, 3$) we denote four-vectors with the notation A^a, B^b, and so on, using a superscript index. The *inner product* between two such vectors is $g_{ab}A^aB^b$ where g_{ab}

is the *metric tensor* of flat Lorentzian space-time, given by the indefinite metric

$$g_{ab} = \begin{pmatrix} 1 & 0 & 0 & 0 \\ 0 & -1 & 0 & 0 \\ 0 & 0 & -1 & 0 \\ 0 & 0 & 0 & -1 \end{pmatrix}. \tag{3.1.1}$$

In the expression $g_{ab}A^aB^b$ there is the usual implied summation over repeated indices; thus more explicitly we have:

$$g_{ab}A^aB^b = A^0B^0 - A^1B^1 - A^2B^2 - A^3B^3.$$

The squared 'magnitude' of a vector A^a is $g_{ab}A^aA^b$, which owing to the indefinite signature of g_{ab} is not necessarily positive. A vector A^a ($\neq 0$) is called *time-like*, *space-like*, or *null* according as to whether $g_{ab}A^aA^b$ is positive, negative, or zero.

A *time-orientation* is chosen by taking at will some time-like vector, say $U^a = (1,0,0,0)$, and designating it to be *future-pointing*. Any other time-like or null vector V^a such that $g_{ab}U^aV^b > 0$ is also future-pointing, whereas if $g_{ab}U^aV^b < 0$ then V^a is *past-pointing*. Space-like vectors are neither future-pointing nor past-pointing.

From these definitions the following general results can be deduced, the proofs of which we leave as an exercise:

(i) if P^a is time-like and $P^aS_a = 0$ then S_a is space-like;
(ii) if P^a and Q^a are time-like and $P^aQ_a > 0$ then either both are future-pointing or both are past-pointing;
(iii) if P^a and Q^a are null and $P^aQ_a = 0$ then P^a and Q^a are proportional;
(iv) if P^a is null and $P^aR_a = 0$ then either R^a is space-like or $R^a \propto P^a$;
(v) if U^a, V^a and W^a are time-like with $U^aV_a > 0$ and $U^aW_a > 0$, then $V^aW_a > 0$.

In the formulae above we have incorporated the following *index-lowering convention*: $P_a = g_{ab}P^b$. Thus for the inner product of P^a and Q^a we may write $P^aQ_a = g_{ab}P^aQ^b = P_bQ^b$. To raise indices we use the inverse metric g^{ab}, defined such that $g^{ab}g_{bc} = \delta^a_c$ where δ^a_b is the Kronecker delta in four dimensions defined by $\delta^a_b = 1$ if $a = b$ and $\delta^a_b = 0$ if $a \neq b$. Then we have $P_a = g_{ab}P^b$, $P^a = g^{ab}P_b$, $P^a = \delta^a_b P^b$, and $P_a = \delta^b_a P_b$.

Two space-time events x^a and y^a are said to be *null-separated* if the expression $g_{ab}(x^a - y^a)(x^b - y^b)$ vanishes, and for a given event x^a the set of all points null-separated from it is called the *light-cone* or *null-cone* at x^a. The null-cone of x^a is generated by a two-dimensional family of *null lines* or *null geodesics* emanating from x^a. Indeed, for any point y^a on the null cone of x^a all points on the line $x^a + \lambda(x^a - y^a)$, λ variable, are null-separated from x^a; this line is the unique *generator* through x^a containing y^a.

In summary we see that the light-cone enables us to classify vectors into five categories: future-pointing time-like, future-pointing null, space-like, past-pointing null, and past-pointing time-like.

3.2   Relativistic kinematics

The relativistic momentum of an idealized classical point-particle can be represented by a four-vector P^a that is required to be future-pointing and time-like (or null, in the case of a particle which moves at the speed of light). The squared *rest mass* of the particle is then given by

$$M^2 = P^a P_a. \tag{3.2.1}$$

If V^a denotes the unit future-pointing time-like four-velocity of an observer, then the total relativistic energy of the particle, measured relative to the observer, is

$$E = P^a V_a. \tag{3.2.2}$$

For example, if the particle is at rest with respect to the space-time coordinates we have chosen (this of course has no *absolute* significance), and if the observer is moving relatively with three-velocity $\mathbf{v}$, then $P^a = (M, 0, 0, 0)$ and $V^a = \gamma(1, \mathbf{v})$ where $\gamma = (1 - v^2)^{-\frac{1}{2}}$ is the *Lorentz factor* associated with the velocity vector $\mathbf{v}$. These conditions ensure $V^a V_a = 1$, and for E we obtain

$$E = M/(1 - v^2)^{\frac{1}{2}}, \tag{3.2.3}$$

a well-known result. Though expression (3.2.3) is very useful for certain purposes, the invariant formula (3.2.2) is the one to bear in mind for future application, where it has a direct general relativistic interpretation.

In particular, if P^a happens to be null then (3.2.3) is no longer applicable, but (3.2.2) is still valid.

Thus if P^a represents the four-momentum of a zero-rest-mass particle (e.g. a photon), then E is the energy of that particle as measured by an observer with unit four-velocity V^a.

Therefore, if a photon with four-momentum P^a is emitted by a body with four-velocity U^a and received by an observer with four-velocity V^a, then the *emitted* energy (measured in the frame of the emitting body) is $P^a U_a$, whereas the energy measured at *reception* is $P^a V_a$. If U^a and V^a differ then so will the two measured energies: this is the *relativistic Doppler effect*.

Thus, for example, if $U^a = (1, 0, 0, 0)$, $V^a = \gamma(1, \mathbf{v})$, and $P^a = (E, \mathbf{p})$ with $E^2 - p^2 = 0$, then the emitted energy is measured by the U^a observer to be

$$P^a U_a = E,$$

whereas the received energy is measured by the V^a observer to be

$$P^a V_a = \gamma E - \gamma \mathbf{p} \cdot \mathbf{v}.$$

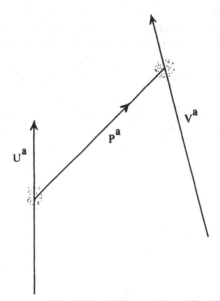

Figure 3.1. Emission and reception of a photon.

Let us denote the emitted energy E_1 and the received energy E_2. Then $E_2/E_1 = P^a V_a/P^a U_a = \gamma(1 - v\cos\theta)$ where θ is the angle between $\mathbf{p}$ and $\mathbf{v}$. Thus, if the velocity of the receiver is in the same direction spatially as the motion of the photon, so $\theta = 0$, then $E_2/E_1 = [(1 - v)/(1 + v)]^{\frac{1}{2}}$ and the received energy is measured to be *less* than the emitted energy. On the other hand, if $\theta = \frac{1}{2}\pi$, corresponding to a receiver whose motion is transverse to that of the photon, then $E_2/E_1 = \gamma$, giving a received energy that is *greater* than that of the emitted energy. Note that in the case of $\theta = 0$ we have $E_2/E_1 \sim 1 - v$ to first order in v, corresponding to the 'classical' red-shift formula; on the other hand when $\theta = \frac{1}{2}\pi$ we have $E_2/E_1 \sim 1$, showing that the transverse 'blue-shift' effect is a purely relativistic one, only making itself felt in higher powers of v. (Bear in mind our units are such that for the speed of light we have $c = 1$.)

When particles interact, the total four-momentum is conserved. Thus if a particle with four-momentum P^a decays into two particles with four-momentum Q_1^a and Q_2^a we must have

$$P^a = Q_1^a + Q_2^a. \qquad (3.2.4)$$

Likewise, if particles with four-momentum P^a and Q^a interact to form new particles with four-momentum $\tilde{P}^a$ and $\tilde{Q}^a$ then we have

$$P^a + Q^a = \tilde{P}^a + \tilde{Q}^a. \qquad (3.2.5)$$

We list below a few examples typical of the conclusions one can draw from these formulae. The applications of this material are very wide indeed.

Example (i). A particle of mass M and four-momentum P^a decays into two massless particles. If we rearrange (3.2.4) to form $P^a - Q_1^a = Q_2^a$, and square each side of this equation, we get $M^2 - 2P^a Q_{1a} = 0$. Since $P^a = MU^a$ where U^a is the time-like unit vector in the direction of P^a it follows that $P^a Q_{1a} = ME_1$ where E_1 is the energy of Q_1^a as measured in the frame of the progenitor. Thus $E_1 = \frac{1}{2}M$ and similarly $E_2 = \frac{1}{2}M$.

For instance, if a neutral π meson (rest mass 135 MeV) decays into two photons it follows that in the rest frame of the meson each photon has energy 67.5 MeV.

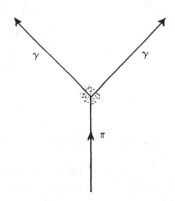

Figure 3.2.i. Decay of a neutral π meson.

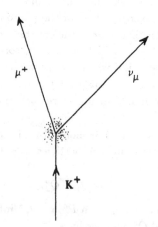

Figure 3.2.ii. Decay of a K^+ meson.

Example (ii). Suppose in (3.2.4) P^a has mass M, whereas $Q_1^a Q_{1a} = m^2$ and $Q_2^a Q_{2a} = 0$. If we square each side of $P^a - Q_2^a = Q_1^a$ it follows that $M^2 - 2ME_2 = m^2$

where E_2 is the energy of Q_2^a in the frame of P^a. Thus $E_2 = (M^2 - m^2)/2M$.

For instance, a K^+ particle decays into an antimuon μ^+ and a muon neutrino. The K^+ and μ^+ particles have masses 494 MeV and 106 MeV respectively and the neutrino is massless. It follows therefore that in the frame of the K^+ particle the neutrino is emitted with energy 236 MeV.

Example (iii). A particle of mass M decays into two particles of masses m_1 and m_2. By squaring $P^a - Q_1^a = Q_2^a$ we get $E_1 = (M^2 + m_1^2 - m_2^2)/2M$, and by squaring $P^a - Q_2^a = Q_1^a$ we get $E_2 = (M^2 - m_1^2 + m_2^2)/2M$.

For instance, a Λ particle (mass 1116 MeV) decays into a proton (mass 938 MeV) and a π^- meson (mass 140 MeV). The proton is therefore emitted with energy 943 MeV, and the π^- meson with energy 173 MeV.

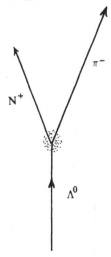

Figure 3.2.iii. Decay of the lambda hyperon.

Example (iv). A photon with four-momentum P^a scatters off an electron with four-momentum Q^a so as to produce a photon of four-momentum $\tilde{P}^a$ and an electron of four-momentum $\tilde{Q}^a$. In the rest frame of the original electron we are to calculate the angle made between the directions of the incident and scattered photons.

The four-momentum of the electron before scattering can be written $Q^a = mV^a$ where V^a is a unit vector. The *three*-momentum of the incident photon, relative to the rest frame, is $P^a - V^a P^b V_b$, which can be written in the form ER^a, where E is the energy of the incident photon in the rest frame, and R^a is the unit space-like vector in the direction of the incident photon's three-momentum.

Similarly, the three-momentum of $\tilde{P}^a$ relative to V^a is $\tilde{P}^a - V^a \tilde{P}^b V_b = \tilde{E}\tilde{R}^a$, where $\tilde{E}$ is the energy of the scattered photon and $\tilde{R}^a$ is the unit space-like vector in the direction of the scattered photon's three-momentum relative to the rest frame.

We have $R^a \tilde{R}_a = -\cos\theta$ for the scattering angle, by vector algebra in three dimensions, the minus sign deriving from the fact that the vectors are space-like. Thus:

$$(P^a - V^a P^b V_b)(\tilde{P}_a - V_a \tilde{P}^b V_b) = -E\tilde{E}\cos\theta,$$

from which we obtain

$$P^a \tilde{P}_a = E\tilde{E}(1 - \cos\theta).$$

By conservation of four-momentum we have $Q^a + P^a - \tilde{P}^a = \tilde{Q}^a$ which when squared gives $2m(E - \tilde{E}) = 2P^a \tilde{P}_a$ since $Q^a Q_a = \tilde{Q}^a \tilde{Q}_a = m^2$, where m is the electron mass. Therefore:

$$E - \tilde{E} = \frac{E\tilde{E}}{m}(1 - \cos\theta).$$

This is the celebrated formula for *Compton scattering*, interrelating the energies of the incident and scattered photons with the scattering angle.

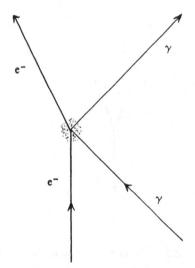

Figure 3.2.iv. Energy transfer via Compton scattering.

If we write $E = h/\lambda$ where h is Planck's constant and λ is the photon wavelength ($c = 1$) then we get

$$\tilde{\lambda} = \lambda + \frac{2h}{m}\sin^2(\theta/2)$$

showing that as θ increases, so does the wavelength of the scattered light relative to a given wavelength.

Examples such as these are abundant in the literature of special relativity, and the reader may find it a useful exercise to translate into the geometry of four-vectors some of the further results that may be found there.

Incidentally, in example (iv) above we have used the fact that if a photon has four-momentum P^a then its three-momentum, relative to the frame defined by a unit time-like vector V^a, is given by $R^a = P^a - V^a P^b V_b$.

More generally, given any vector P^a we can *project* it into the space-like hyperplane (3-space) orthogonal to V^a, and the resulting projection is defined by R^a. It is a straightforward matter to verify that $R^a V_a = 0$, showing that R^a lies on the space of vectors orthogonal to V^a.

Alternatively, we can write $P^a = V^a (P^b V_b) + R^a$, splitting P^a into its *longitudinal* and *transverse* components relative to V^a. The *projection operator* Π_b^a associated with V^a is given by

$$\Pi_b^a = \delta_b^a - V^a V_b$$

and it is easy to see that $\Pi_b^a P^b = R^a$, and that $\Pi_b^a \Pi_c^b = \Pi_c^a$.

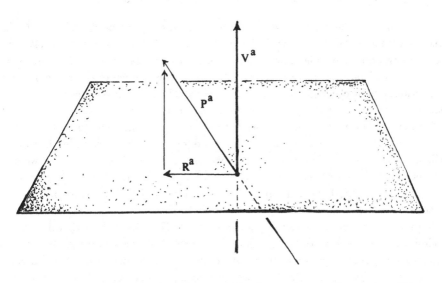

Figure 3.3. Projection into the space-like hyperplane orthogonal to V^a.

3.3 Proper time

When idealized as a point particle, a massive body unacted upon by gravitation or external forces has as its trajectory or *world-line* in space-time a straight time-like line. We refer to such paths equivalently as *time-like geodesics* or *inertial trajectories*, the relevant formula being:

$$x^a(s) = b^a + s V^a \qquad (3.3.1)$$

where b^a is a point on the world-line, and V^a is a unit future-pointing time-like vector in the direction of the world-line. When a geodesic is represented in this way, the parameter s is the *proper time* along the trajectory. The difference $s_2 - s_1$ for two points $x^a(s_2)$ and $x^a(s_1)$ on the trajectory is precisely the physical time lapse measured by a 'little man' moving from one event to the other along the trajectory.

As a geometrical entity (i.e. as a point set in space-time) the geodesic can be parametrized in ways other than (3.3.1): for example,

$$x^a(\sigma) = b^a + g(\sigma)V^a \tag{3.3.2}$$

represents the same geodesic, provided $g : \sigma \to (-\infty, \infty)$. What distinguishes (3.3.1) from (3.3.2) is the *normalization condition* $\dot{x}^a \dot{x}_a = 1$ where the dot denotes d/ds. If we adopt this normalization then the equation for a time-like geodesic is $\ddot{x}^a = 0$, the general solution of which is of the form (3.3.1).

3.4   Lorentz contraction and time dilation

These remarkable phenomena are usually introduced early in the discussion of relativistic kinematics, side by side with the notion of *Lorentz transformation*. Relativistic length contraction and time dilation can be accounted for satisfactorily, nevertheless, on the basis of pure geometric considerations, without direct reference to Lorentz transformations or other group theoretic notions. The relevant constructions are, in some respects, more transparent when presented in this way.

The Lorentz contraction effect arises out of a subtle ambiguity as to what is meant by the 'length' of an extended body. If we think of the body as extended in time as well, so as to form a kind of space-time entity analogous to a world line, then we may take space-like *cross-sections* of this entity in various ways with space-like hyperplanes; and the relevant length calculations vary in result according as to which hyperplane is chosen.

Consider, for example, a pair of parallel time-like geodesics. We may think of the geodesics as representing the endpoints of a rigid rod not acted on by any external force. Now suppose an observer based at point P has unit four-velocity U^a, parallel to the two time-like geodesics. The space-like hyperplane orthogonal to U^a intersects the geodesics at a pair of points α and β. Therefore, relative to the frame of reference defined by U^a the rod is represented by the vector $R^a = \overrightarrow{\alpha\beta}$, which has length $L^2 = -R^a R_a$. This is evidently also the length of the rod in its own rest frame, which coincides by construction with that of U^a. Thus we may call L the *proper length* of the rod.

On the other hand, suppose we consider an observer based at Q with four-velocity V^a. The space-like hyperplane through Q orthogonal to V^a intersects the two geodesics at the points ξ and ζ, and thus in the frame of V^a the rod is represented by the vector $\tilde{R}^a = \overrightarrow{\xi\zeta}$, which has squared length $\tilde{L} = -\tilde{R}^a \tilde{R}_a$.

Now by elementary geometry, it is evident that $\tilde{R}^a = R^a + \psi U^a$ for some ψ, and since $\tilde{R}^a V_a$ must vanish we have $\psi = -R^a V_a / U^b V_b$. It follows therefore that

$$\tilde{L}^2 = -\tilde{R}^a \tilde{R}_a$$

$$= -(R^a + \psi U^a)(R_a + \psi U_a)$$
$$= -R^a R_a - \psi^2 U^a U_a$$
$$= L^2 - (R^a V_a / U^b V_b)^2 \tag{3.4.1}$$

by use of $R^a U_a = 0$. In all circumstances we have $\tilde{L} < L$, showing in general that the length measured in the frame V^a will be less than the proper length of the rod. This is the *Lorentz contraction effect*, and formula (3.4.1) gives a precise measure of the effect for an arbitrary observer with four-velocity V^a moving relative to a rod with position R_a relative to its own four-velocity U^a.

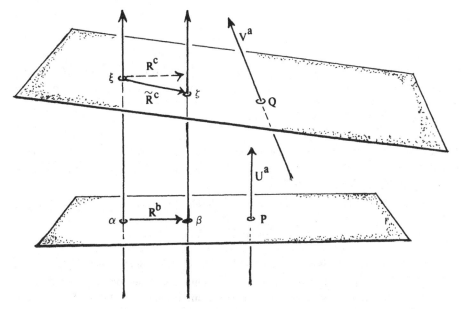

Figure 3.4. Lorentz contraction of the rod $\overrightarrow{\alpha\beta}$.

Let us now specialize further to the case in which the rod is pointed in the same direction spatially as the motion of V^a relative to U^a. Then we have $R^a = LX^a$ where X^a is a unit space-like vector, and

$$V^a = \gamma[U^a + vX^a] \tag{3.4.2}$$

where v is the speed and $\gamma = (1 - v^2)^{-1/2}$, so $V^a V_a = 1$. We easily calculate $R^a V_a = \gamma v L$ and $U^a V_a = \gamma$, from which it follows from (3.4.1) that

$$\tilde{L} = (1 - v^2)^{1/2} L, \tag{3.4.3}$$

Einstein's famous formula for this effect. Formula (3.4.1) is applicable to the slightly more general circumstances where the rod is not necessarily oriented in any special way with respect to the moving observer.

Note that it is the *space-like hyperplane*, through a given point, orthogonal to the four-velocity of the 'observer', that defines that observer's concept of *simultaneity*. Any two events lying in the same space-like hyperplane orthogonal to an observer's world-line are regarded as 'simultaneous' by that observer.

Bearing this in mind, let us now turn to the phenomenon of *time-dilation*. Two world-lines cross at point A, one with four-velocity U^a and the other with four-velocity V^a. After passage of a specified duration of proper time $\tilde{\tau}$ the second observer reaches the point $\tilde{B}$. This point is simultaneous, in the frame of the first observer, with a point B on the first observer's world-line. The space-like hyperplane Π through B and orthogonal to U^a intersects the second world-line at $\tilde{B}$.

By elementary geometry we have

$$\tau U^a + r X^a = \tilde{\tau} V^a \qquad (3.4.4)$$

for some value of r where X^a is a unit space-like vector in the direction $B\tilde{B}$, and τ is the proper time elapsed along the trajectory AB. By virtue of (3.4.2) we have $r = \gamma v \tilde{\tau}$ and

$$\tilde{\tau} = (1 - v^2)^{1/2} \tau. \qquad (3.4.5)$$

Thus, while time $\tilde{\tau}$ elapses along the trajectory $A\tilde{B}$ the larger quantity $\tau = \tilde{\tau}\gamma$ elapses along the trajectory AB, so that *the first observer will claim the second observer's clock is running slow relative to his.*

The relationship between U^a and V^a is completely symmetrical, since (3.4.2) implies

$$U^a = \gamma[V^a - v\tilde{X}^a] \qquad (3.4.6)$$

where $\tilde{X} = X^a - \gamma v V^a$ is a unit space-like vector orthogonal to V^a. Thus, in the frame of V^a when proper time $\tilde{\tau}$ has elapsed along the trajectory $A\tilde{B}$, we find that the first observer has moved only as far as the point B', with an elapse of proper time τ' such that

$$\tau' U^a + r' \tilde{X}^a = \tilde{\tau} V^a. \qquad (3.4.7)$$

By a comparison of formulae (3.4.6) and (3.4.7) we obtain

$$\tau' = (1 - v^2)^{1/2} \tilde{\tau} \qquad (3.4.8)$$

showing that relative to his own frame the *second* observer will think the *first* observer's clock is running slow.

It will seem slightly surprising that the hyperplane $\tilde{\Pi}$ intersects the world-line AB at a point B' earlier than B, rather than later as naive intuition might suggest. But this is a manifestation of the hyperbolic signature of the metric, and also illustrates that some care is required in the interpretation of figures drawn to represent relativistic phenomena.

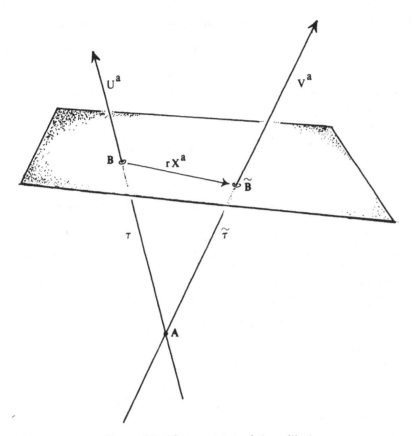

Figure 3.5. The geometry of time dilation.

3.5   Relativistic electrodynamics

Maxwell's equations lend themselves very naturally to a relativistic formulation—
so much so that it can be argued that Einstein was in some respects anticipated by
Maxwell and his followers. In any case it was in part through a careful consideration
of electromagnetic phenomena that Einstein was led to special relativity.

The electromagnetic field is represented in relativistic terms by a second rank
anti-symmetric tensor F_{ab} $(a, b = 0, 1, 2, 3)$ that satisfies $F_{ab} = -F_{ba}$ and has as
its components the following non-vanishing elements:

$$E_i = F_{i0} \quad (i = 1, 2, 3) \tag{3.5.1}$$

$$B_i = \frac{1}{2}\varepsilon_{ijk}F_{jk}. \tag{3.5.2}$$

Here, in order to compare the relativistic and non-relativistic theories, we use a convenient mixture of three and four dimensional notation. Since $F_{ij} = -F_{ji}$ we may

invert (3.5.2) by use of (2.1.1) in order to obtain

$$F_{ij} = \varepsilon_{ijk} B_k. \tag{3.5.3}$$

Thus by use of (3.5.1), (3.5.2) and (3.5.3) we may go back and forth from the electric and magnetic field vectors to the electromagnetic field tensor.

If we introduce the space-time derivative operator $\nabla_a = \partial/\partial x^a$ then Maxwell's vacuum equations are expressed as follows:

$$\nabla^a F_{ab} = 0 \tag{3.5.4}$$

$$\nabla_a F_{bc} + \nabla_b F_{ca} + \nabla_c F_{ab} = 0 \tag{3.5.5}$$

where $\nabla^a = g^{ab}\nabla_b$. Indeed, if we expand these equations out relative to the coordinate system (t, x_i) we obtain

$$\nabla^i F_{i0} = 0 \tag{3.5.6}$$

$$\nabla^0 F_{0j} + \nabla^i F_{ij} = 0 \tag{3.5.7}$$

for (3.5.4); and

$$\nabla_i F_{jk} + \nabla_j F_{ki} + \nabla_k F_{ij} = 0 \tag{3.5.8}$$

$$\nabla_0 F_{ij} + \nabla_i F_{j0} + \nabla_j F_{0i} = 0 \tag{3.5.9}$$

for (3.5.5). From (3.5.6) we obtain

$$\nabla_i E_i = 0. \tag{3.5.10}$$

Since $\nabla^i F_{ij} = -\nabla_i F_{ij} = -\nabla_i \varepsilon_{ijk} B_k = \varepsilon_{jpq} \nabla_p B_q$, and since $\nabla^0 F_{0j} = \nabla_0 F_{0j} = \dot{E}_j$, it follows from (3.5.7) that

$$\varepsilon_{jpq} \nabla_p B_q = \dot{E}_j. \tag{3.5.11}$$

By contraction of (3.5.8) with ε_{ijk} we get

$$\nabla_i B_i = 0 \tag{3.5.12}$$

after use of (3.5.2). And finally by contraction of (3.5.9) with $\frac{1}{2}\varepsilon_{kij}$ we obtain

$$\dot{B}_k + \varepsilon_{kij} \nabla_i E_j = 0. \tag{3.5.13}$$

We recognise (3.5.10) to (3.5.13) as familiar three-dimensional expressions for Maxwell's equations (cf. equations 2.4.1 to 2.4.4)

Thus Maxwell's equations in vacuo are given relativistically by (3.5.4) and (3.5.5). If sources are present then (3.5.5) remains unchanged, but (3.5.4) is modified so as to read

$$\nabla^a F_{ab} = -4\pi J_b \tag{3.5.14}$$

where J^a is the *electromagnetic four-current*, given in our special coordinate system by

$$J^a = (\rho, \mathbf{J}).$$

A calculation then shows that (3.5.14) leads to (2.4.2) and (2.4.4). Moreover since F_{ab} is anti-symmetric it follows from (3.5.14) that $\nabla_a J^a = 0$, which is equivalent to the conservation law (2.4.7).

If we contract (3.5.5) with ∇^a and use (3.5.14) then after some rearrangement we obtain

$$\Box F_{ab} = -4\pi(\nabla_a J_b - \nabla_b J_a)$$

where $\Box = \nabla^a \nabla_a$ is the *wave operator*. It follows at once that if $J_a = 0$ then each component of the electromagnetic field tensor satisfies the wave equation; but a necessary and sufficient condition for this to be the case is evidently merely that the current should be curl-free, viz. of the form $J_a = \nabla_a \chi$ for some scalar χ.

Equation (3.5.5), which holds whether or not sources are present, is a necessary and sufficient condition locally for there to exist a *vector potential* A_b such that

$$F_{ab} = \nabla_a A_b - \nabla_b A_a.$$

Indeed it follows at once that $F_{ij} = \nabla_i A_j - \nabla_j A_i$ which by virtue of (3.5.2) gives (2.4.5); and from $F_{i0} = \nabla_i A_0 - \nabla_0 A_i$ we obtain (2.4.6) by virtue of (3.5.1) with the identification $A_0 = \phi$.

From the equation (3.5.14) it follows that the vector potential satisfies

$$\nabla_b \nabla^b A_a = -4\pi J_a + \nabla_a(\nabla_b A^b).$$

However, since A_a is determined only up to a transformation of the form $A_a \rightarrow A_a + \nabla_a \psi$, under which F_{ab} is invariant, we may choose ψ so as to make $\nabla_b A^b$ constant so $\nabla_b \nabla^b A_a = -4\pi J_a$.

At this point it is convenient to introduce some further notational refinements. For the *anti-symmetric* product of two vectorial quantities A_a and B_b we write

$$A_{[a} B_{b]} = \frac{1}{2}(A_a B_b - A_b B_a).$$

Round brackets are used for the *symmetric* product:

$$A_{(a} B_{b)} = \frac{1}{2}(A_a B_b + A_b B_a).$$

Note therefore the identity

$$A_a B_b = A_{[a} B_{b]} + A_{(a} B_{b)}.$$

More generally for any tensor C_{ab} we write

$$C_{ab} = C_{[ab]} + C_{(ab)}$$

where

$$C_{[ab]} = \frac{1}{2}(C_{ab} - C_{ba})$$

and

$$C_{(ab)} = \frac{1}{2}(C_{ab} + C_{ba}).$$

We call $C_{[ab]}$ and $C_{(ab)}$ the anti-symmetric and symmetric parts of C_{ab}, respectively. Furthermore in the case of three-index tensors we write

$$D_{(abc)} = \frac{1}{6}(D_{abc} + D_{bca} + D_{cab} + D_{bac} + D_{cba} + D_{acb})$$

and

$$D_{[abc]} = \frac{1}{6}(D_{abc} + D_{bca} + D_{cab} - D_{bac} - D_{cba} - D_{acb}).$$

Note that if F_{bc} is itself anti-symmetric then

$$E_{[a}F_{bc]} = \frac{1}{3}(E_aF_{bc} + E_bF_{ca} + E_cF_{ab}).$$

With this notation the Maxwell equation (3.5.5) may be written neatly as $\nabla_{[a}F_{bc]} = 0$, which implies by (3.5.13) the existence of a vector A_c such that $F_{bc} = 2\nabla_{[b}A_{c]}$. Similarly, equation (3.5.12) may be written $\nabla_c\nabla^cF_{ab} = -8\pi\nabla_{[a}J_{b]}$.

The *stress-energy tensor* of the electromagnetic field is a symmetric second rank tensor T_{ab} defined as follows:

$$4\pi T_{ab} = F_{ac}F_{bd}g^{cd} - \frac{1}{4}g_{ab}F_{de}F^{de}. \qquad (3.5.15)$$

Here we have an interesting example of the sort of economy that arises in relativistic expressions, since T_{ab} embodies in a single tensor field the various expressions for electromagnetic energy density, energy flux and stress considered in section 2.4. In a vacuum (i.e. for source-free fields) T_{ab} has *vanishing divergence*,

$$\nabla^aT_{ab} = 0, \qquad (3.5.16)$$

expressing the conservation of electromagnetic energy and momentum. To see that (3.5.16) holds, consider the term $F_{ac}F_{bd}g^{cd}$ in (3.5.15). If we take its divergence we obtain

$$\begin{aligned}
\nabla^a(F_{ac}F_{bd}g^{cd}) &= F_{ac}\nabla^aF_{bd}g^{cd} \quad \text{(by 3.5.4)} \\
&= F^{ad}\nabla_aF_{bd} \\
&= -F^{ad}(\nabla_bF_{da} + \nabla_dF_{ab}) \quad \text{(by 3.5.5)} \\
&= \frac{1}{2}\nabla_b(F_{de}F^{de}) - F^{ad}\nabla_dF_{ab} \\
&= \frac{1}{2}\nabla_b(F_{de}F^{de}) - \nabla^a(F_{ac}F_{bd}g^{cd}).
\end{aligned}$$

from which it follows that

$$\nabla^a(F_{ac}F_{bd}g^{cd}) = \frac{1}{4}\nabla_b(F_{de}F^{de})$$

which leads immediately to (3.5.16). The stress-energy tensor plays an important part in the development of relativistic theory.

As a further notational development we introduce the *totally skew tensor* ε_{abcd} in four dimensions defined by

$$\varepsilon_{abcd} = \begin{cases} 1 & \text{if } abcd \text{ is an even permutation of 0123} \\ -1 & \text{if } abcd \text{ is an odd permutation of 0123} \\ 0 & \text{otherwise.} \end{cases}$$

It satisfies the following identities:

$$\varepsilon_{abcd}\varepsilon^{abcd} = -24 \qquad (3.5.17)$$

$$\varepsilon_{abcd}\varepsilon^{abch} = -6\delta_d^h \qquad (3.5.18)$$

$$\varepsilon_{abcd}\varepsilon^{abgh} = -4\delta_{[c}^{[g}\delta_{d]}^{h]} \qquad (3.5.19)$$

$$\varepsilon_{abcd}\varepsilon^{afgh} = -6\delta_{[b}^{[f}\delta_c^g\delta_{d]}^{h]} \qquad (3.5.20)$$

$$\varepsilon_{abcd}\varepsilon^{efgh} = -24\delta_{[a}^{[e}\delta_b^f\delta_c^g\delta_{d]}^{h]} \qquad (3.5.21)$$

where $\varepsilon^{abcd} = g^{ap}g^{bq}g^{cr}g^{ds}\varepsilon_{pqrs}$. Note that $\varepsilon^{0123} = -\varepsilon_{0123}$. The *dual* of an anti-symmetric tensor F_{ab} is defined by

$$^*F_{ab} = \frac{1}{2}g_{ap}g_{bq}\varepsilon^{pqrs}F_{rs} \qquad (3.5.22)$$

or more simply,

$$^*F_{ab} = \frac{1}{2}\varepsilon_{abcd}F^{cd}. \qquad (3.5.23)$$

By virtue of (3.5.19) we have the identity

$$^{**}F_{ab} = -F_{ab}. \qquad (3.5.24)$$

The effect of the duality operation on an electromagnetic field, for example, is to replace E_i with B_i, and B_i with $-E_i$. Repetition of this procedure then evidently leads to (3.5.24).

In particular, if we take the second Maxwell equation (3.5.5) in the form $\nabla_{[a}F_{bc]} = 0$ and contract with ε^{abcd}, then by (3.5.23) we obtain:

$$\nabla^{a*}F_{ab} = 0 \Leftrightarrow \nabla_{[a}F_{bc]} = 0. \qquad (3.5.25)$$

For certain purposes it is useful to produce a pair of complex tensors $^+F_{ab}$ and $^-F_{ab}$ called the *self-dual* and *anti-self-dual* parts of F_{ab} defined by

$$^+F_{ab} = F_{ab} - i^*F_{ab}, \qquad (3.5.26)$$

$$^-F_{ab} = F_{ab} + i^*F_{ab}, \qquad (3.5.27)$$

so called because under duality they satisfy

$$^{*+}F_{ab} = i^+F_{ab},$$

$$^{*-}F_{ab} = -i^-F_{ab}. \qquad (3.5.29)$$

By virtue of (3.5.4) and (3.5.25) we see that the vacuum equations are expressed neatly by the single relation

$$\nabla_a{}^+F^{ab} = 0 \qquad (3.5.30)$$

or, equivalently, by $\nabla_{[a}{}^+F_{bc]} = 0$, $\nabla_a{}^-F^{ab} = 0$, or $\nabla_{[a}{}^-F_{bc]} = 0$.

Let us turn now to consider the equations of motion of a charged particle under the influence of an electromagnetic field F_{ab}. Let us suppose the trajectory of the particle is given by a time-like curve $x^a(s)$ normalized so that $\dot{x}^a \dot{x}_a = 1$, and that the particle has mass m and charge q. The equations of motion are given by the *Lorentz equation*

$$m\ddot{x}^a = q F^a{}_b \dot{x}^b \tag{3.5.31}$$

where the field $F^{ab}(x^c(s))$ is evaluated, for each value of s, at the appropriate point along the trajectory. Clearly when the electric charge q vanishes motion is along a geodesic. Note that since F^{ab} is anti-symmetric we have $\dot{x}_a \ddot{x}^a = 0$, showing that (3.5.31) is consistent with the normalization condition we have imposed on $\dot{x}^a$.

3.6   Relativistic hydrodynamics

An ideal or perfect *relativistic fluid* is characterized by its energy density ρ, its pressure p, and its velocity field u^a, which is required to be everywhere time-like and normalized so that $u^a u_a = 1$. The equations of motion of the fluid are equivalent to the vanishing of the divergence of its stress-energy tensor T_{ab} which is defined by

$$T_{ab} = (\rho + p)u_a u_b - g_{ab} p. \tag{3.6.1}$$

It is a straightforward matter that $\nabla^a T_{ab} = 0$ implies the *relativistic Euler equation*

$$(\rho + p)u^a \nabla_a u^b = (g^{bc} - u^b u^c)\nabla_c p, \tag{3.6.2}$$

and the *relativistic continuity equation*

$$u^a \nabla_a \rho + (\rho + p)\nabla_a u^a = 0. \tag{3.6.3}$$

These relations are the relativistic analogues of non-relativistic equations (2.5.1) and (2.5.2) respectively. For certain purposes it useful to write (3.6.2) in the slightly modified form

$$(\rho + p)u^a \nabla_a u^b = \Pi^{bc} \nabla_c p \tag{3.6.4}$$

so as to emphasize the role of the projection operator (cf. § 3.2)

$$\Pi^{bc} = g^{bc} - u^b u^c \tag{3.6.5}$$

which at each point in space-time projects the pressure gradient $\nabla_c p$ into the space-like hyperplane orthogonal to u^a at that point.

   To see that $\nabla^a T_{ab} = 0$ does indeed lead to (3.6.2) and (3.6.3) we proceed as follows. An elementary calculation, if we bear in mind that $u^a u_a = 1$, gives

$$\nabla^a T_{ab} = (\rho + p)u_a \nabla^a u_b + \nabla^a[(\rho + p)u_a]u_b - \nabla_b p, \tag{3.6.6}$$

whence $u^b \nabla^a T_{ab} = 0$ implies

$$\nabla^a[(\rho + p)u_a] - u^b \nabla_b p = 0 \tag{3.6.7}$$

which after some rearrangement gives (3.6.3). Then if $\nabla^a[(\rho + p)u_a] = u^b \nabla_b p$ is inserted back into (3.6.6) we are lead directly to (3.6.2).

These relations, as we shall see, can be reformulated readily in the case of general relativity, where they have numerous important applications. Generally it is necessary to provide further information so as to characterize the nature of the particular fluid substance under consideration.

The most straightforward case is that of a *simple* perfect fluid, which is to be regarded as composed of particles of a single type with number density $n(x^a)$. For example, $n(x^a)$ might be the number of hydrogen atoms per cubic centimetre. The ratio ρ/n is then the *specific energy*, i.e. the total energy per particle at the point x^a. We supplement the variables n, ρ and p with two *thermodynamic* variables—the *specific entropy* S (entropy per particle at the point x^a), and the *temperature* T. These are related to n, ρ and p by the laws of thermodynamics. In particular, we have the key thermodynamic relation

$$kTdS = d(\rho/n) + pd(1/n) \qquad (3.6.8)$$

where k is Boltzmann's constant ($1.03806 \times 10^{-16} \ erg/deg$), ρ/n is the specific energy, and $1/n$ is the 'specific volume' (i.e. volume per particle at the point x^a).

Equation (3.6.8) is to be thought of as a 'local' formulation of the classical thermodynamic law $kTd\Sigma = dU + pdV$, where Σ is the total entropy. After rearrangement, it can be put into the form

$$dS = \frac{1}{knT}d\rho - \frac{(\rho + p)}{kn^2T}dn. \qquad (3.6.9)$$

Thus if we regard ρ and n as independent variables on which S, T and p depend, then we have

$$\frac{\partial S}{\partial \rho} = \frac{1}{knT} \qquad (3.6.10)$$

and

$$\frac{\partial S}{\partial n} = -\frac{1}{kn^2T}(\rho + p). \qquad (3.6.11)$$

By a fundamental *complete equation of state* for a simple perfect fluid we mean a specification of the entropy function $S(\rho, n)$. Indeed, once the specific entropy has been given as a function of energy density and number density then equations (3.6.10) and (3.6.11) determine T and p as functions of ρ and n.

For a simple fluid we also require the particle number density n to satisfy the local conservation law

$$\nabla_a(nu^a) = 0. \qquad (3.6.12)$$

It is certainly legitimate to consider fluids in which (3.6.12) does not hold, but then the situation is a rather more complicated one, where the hydrodynamic and thermodynamic relations must be supplemented by further relations characterizing the creation and annihilation of constituent particles.

In summary, then, the motion of a simple fluid is characterized relativistically by the Euler equation (3.6.1), the continuity equation (3.6.3), an equation of state

$S = S(\rho, n)$, the two thermodynamic relations (3.6.10) and (3.6.11), and the particle number conservation relation (3.6.12). To these we must add appropriate initial conditions and boundary conditions.

Now what about the second law of thermodynamics? A perfect fluid is 'perfect' precisely in the sense that there are no viscous or dissipative interactions present amongst its constituents—and as a consequence the total entropy *must* remain constant, and indeed the specific entropy must be constant along the trajectory of any constituent of the fluid since there is no process whereby it might lessen its own entropy at the cost of increasing the entropy of a neighbouring constituent.

In order to see that this is indeed the case we write (3.6.9) in the form

$$knT\nabla_a S = \nabla_a \rho - f\nabla_a n \qquad (3.6.13)$$

where $f = (\rho + p)/n$ is the *specific enthalpy* of the fluid. Contraction of this relation with u^a gives $knTu^a\nabla_a S = u^a\nabla_a \rho - fu^a\nabla_a n$, which by use of the continuity equation (3.6.3) in the form $u^a\nabla_a \rho + fn\nabla_a u^a = 0$ gives

$$knTu^a\nabla_a S = -fn\nabla_a u^a - fu^a\nabla_a n$$
$$= -f\nabla_a(nu^a) \qquad (3.6.14)$$

and thus by virtue of (3.6.12) we see that $u^a\nabla_a S = 0$, showing that S is conserved along the trajectories of the fluid constituents. For a simple perfect fluid we say therefore that the motion is *locally adiabatic*.

This sort of reasoning can be taken much further. For the moment it suffices to note that the principal *dynamical* equations are summarized in the conservation of the relativistic stress-energy tensor; whereas the remaining relations are of a particular character, varying according to details of the problem at hand.

Exercises for chapter 3

[3.1] Show that the sum of two future-pointing null vectors is a future-pointing time-like vector, except when the two null vectors have the same direction. Conversely, show that any time-like vector can be expressed as a sum of two null vectors. For a given time-like vector the two null vectors are not uniquely determined: what is the nature of the freedom in their choice?

[3.2] A particle with mass M decays into two particles of masses m_1 and m_2. In the rest frame of M show that the three-momentum of m_1 is of the following magnitude:

$$|\mathbf{P_1}| = [\frac{[M^2 - (m_1 + m_2)^2][M^2 - (m_1 - m_2)^2]}{4M^2}]^{1/2}.$$

[3.3] An eccentric Oxford professor builds a time machine in his sitting room. It consists of a device that propels his leather armchair around a circular track (one metre in radius) at very high speed. How many revolutions per second must the machine make in order for an entire term (eight weeks) to pass by in what appears to be one minute on the professor's watch as he sits in the armchair? (It is assumed that the professor is of exceptional physical fitness!)

[3.4] Suppose F_{ab} satisfies $F_{ab} = F_{[ab]}$ and $F_{[ab}F_{c]d} = 0$. Show that there exist vectors U_a and V_a such that $F_{ab} = U_{[a}V_{b]}$.

[3.5] Let t^a be a future-pointing constant vector with unit norm ($t^a t_a = 1$). Define

$$r^a = x^a - t^a(t^b x_b), \quad r = (-r^a r_a)^{1/2}.$$

Show that if $A^a = er^{-1}t^a$ (e constant) then $F_{ab} = \nabla_{[a}A_{b]}$ satisfies Maxwell's equations with $J_a = 0$ (except at $r = 0$). Suppose that in the definition of r^a we replace x^a by the *complex point* $x^a + ib^a$, with b^a real and space-like; show that F_{ab} continues to satisfy equations (3.5.4) and (3.5.5). Where is the real part of F_{ab} singular?

[3.6] Suppose that F_{ab} satisfies Maxwell's vacuum equations. Verify that the *stress-energy tensor*

$$T_{ab} = F_{ac}F_{bd}g^{cd} - \frac{1}{4}g_{ab}F_{cd}F^{cd}$$

is then necessarily 'conserved', i.e. $\nabla^a T_{ab} = 0$. Conversely, show that if $\nabla^a T_{ab} = 0$ and if $det(F_{ab}) \neq 0$ then if F_{ab} satisfies (3.5.5) it necessarily satisfies (3.5.4). Suppose that J_a is either time-like or else vanishes. Show then that $\nabla^a T_{ab} = 0$ and (3.5.5) imply (3.5.4).

[3.7] With a stress tensor T_{ab} defined as in exercise [3.6] above, show that $T_{ab} = k_a k_b$ for some vector k_a if and only if $F_{ab}F^{ab} = 0$ and $^*F_{ab}F^{ab} = 0$ (show also that these two conditions are equivalent to $\mathbf{E}^2 = \mathbf{B}^2$ and $\mathbf{E} \cdot \mathbf{B} = 0$). Verify that $\nabla^a T_{ab} = 0$ implies k^a is geodesic, i.e. that the vector field k^a satisfies $k^a \nabla_a k^b = \lambda k^b$ for some λ.

[3.8] Show that in Minkowski space the general solution of the equation $\nabla_{(a}\xi_{b)} = 0$ (*Killing's equation*) is $\xi_b = M_{bc}x^c + P_b$ where M_{bc} and P_b are constant and $M_{bc} = M_{[bc]}$.

[3.9] Find the general solution of the second rank 'Killing tensor' equation $\nabla_{(a}K_{bc)} = 0$ in Minkowski space. On how many parameters does K_{ab} depend?

[3.10] Verify equation (3.6.13). Show that for *isentropic* flow (S = constant) Euler's equations for a simple fluid reduce to

$$u^a \nabla_a u^b = (g^{ab} - u^a u^b) \nabla_a \ell n(f),$$

where f is the *specific enthalpy* function $(\rho + p)/n$.

[3.11] For general *locally adiabatic* flow ($u^a \nabla_a S = 0$) show that Euler's equations for a simple fluid are equivalent to

$$C^a \Omega_{ab} = -\frac{1}{2} T f \nabla_b S,$$

where $C^a = fu^a$, and $\Omega_{ab} = \nabla_{[a} C_{b]}$.

[3.12] Suppose $f^i(M^{ab}, N^c)$ comprises three functions ($i = 1, 2, 3$) of M^{ab} and N^c, where M^{ab} is anti-symmetric. A vector field $v^b(x^a)$ satisfies

$$f^i(x^{[a} v^{b]}, v^c) = 0.$$

Show that $v^b(x^a)$ satisfies the *geodesic* equation

$$v^a \nabla_a v^b = \lambda v^b$$

for some scalar $\lambda(x)$.

[3.13] An *isolated massive system* in special relativity is defined by a total momentum P^a and a total angular momentum $M_{(0)}^{ab}$ (anti-symmetric). The angular momentum is specified with respect to an origin O in space-time. Suppose the position vector of a point A (with respect to O) is A_b. Then the angular momentum about A is

$$M_{(A)}^{ab} = M_{(0)}^{ab} - 2A^{[a} P^{b]}.$$

Three of the components of M^{ab} correspond to the classical non-relativistic angular momentum—the other three determine the *centre of mass* of the system. Show that the points X^a which satisfy

$$M_{(x)}^{ab} P_b = 0$$

are of the form

$$X^a = M^{-2} M_{(0)}^{ab} P_b + \lambda P^a$$

where $M^2 = P^a P_a$ and λ is an arbitrary parameter. We thus obtain a geodesic $X^a(\lambda)$ with tangent vector P^a. This is the *relativistic centre of mass* of the isolated system.

[3.14] Consider the eigenvalue equation $F_{ab} V^b = \lambda V_a$ where F_{ab} is the electromagnetic field tensor. Show that any eigenvector satisfying this equation with non-zero eigenvalue is a *null vector*. By considering the expression for the Lorentz force, or otherwise, show that λ satisfies

$$\lambda^4 - \lambda^2(E^2 - B^2) - (E \cdot B)^2 = 0.$$

Deduce that if $E \cdot B \neq 0$ then there exists a pair of unit orthogonal space-like vectors X^a and Y^a such that

$$F_{ab}X^b = \mu Y_a, \quad F_{ab}Y^a = -\mu X_a$$

for a positive real number μ, and a pair of null vectors L^a and N^a such that

$$F_{ab}L^b = \lambda L^a, \quad F_{ab}N^b = -\lambda N^a$$

for a positive real number λ.

[3.15] Given a fundamental equation of state $S = S(\rho, n)$ show that the pressure p and the temperature T of a simple ideal fluid can be expressed in terms of the energy density ρ and the particle density n as follows:

$$p = -\rho - n(S_n/S_\rho), \quad T = 1/(nS_\rho)$$

where $S_\rho = \partial S/\partial \rho$ and $S_n = \partial S/\partial n$. Show that the pressure vanishes if and only if S depends on the specific energy ρ/n alone. Show that the pressure can be expressed as a function $p = P(\rho)$ of the energy density alone if and only if $S(\rho, n) = \sigma(x)$ with $x = \rho g(\rho)/n$ where $g(\rho)$ is an arbitrary function. How is $g(\rho)$ related to $P(\rho)$?

4   Tensor analysis on manifolds

'I have become imbued with great respect for mathematics, the subtler parts of which I had in my simplemindedness regarded as pure luxury until now.'

—Albert Einstein (letter to Arnold Sommerfeld, 1916)

4.1   Informal definition of a manifold

FOR OUR MODEL of space-time we want a topological space that looks 'locally' like $\mathbf{R}^4$ (Euclidean four-space), and on which the familiar operations of calculus are in some sense applicable. This leads us to the notion of a *differentiable manifold*, which we now proceed to describe in the more general context of an n-dimensional setting.

An n-dimensional *topological manifold* is a Hausdorff topological space M with the property that every point in M has an open neighbourhood homeomorphic to an open set in $\mathbf{R}^n$.

Thus we take M to be covered by a family of open sets U_i such that each open set of M can be mapped to an open set of $\mathbf{R}^n$. The sets U_i are called *coordinate patches*. If we regard a point $p \in M$ as belonging to a particular open set (say U_1) in M then its image in $\mathbf{R}^n$ under the map $U_1 \xrightarrow{x^a} \mathbf{R}^n$ is a set of n numbers $x^a[p]$ called the *coordinates* of p (relative to the patch U_1.)

Let us write $U_{ij} = U_i \cap U_j$ for the various intersection regions of the patches. Evidently for points lying in the intersection regions we can assign more than one set of coordinates. Thus we require that there should exist a set of *coordinate transition functions* associated with the overlap regions.

Suppose a point p lies in the overlap region U_{12} and $U_1 \xrightarrow{x^a} \mathbf{R}^n$, $U_2 \xrightarrow{y^a} \mathbf{R}^n$. Then $p \rightarrow x^a[p]$ when p is regarded as in U_1, and $p \rightarrow y^a[p]$ when p is regarded as in U_2. For all p in U_{12} we require the existence of a set of functions $x^a(y^b)$ such that $x^a[p] = x^a(y^b[p])$. (No confusion should result here over the fact that we use the symbol x^a both to denote the coordinate of the point p in the U_1 patch, as well as to denote the associated coordinate transition function—if desired a slightly less ambiguous notation can be employed such as $x^a[p] = f_{12}^a(y^b[p])$ where f_{12}^a denotes the coordinate transition function in the U_{12} patch.)

We require the coordinate transition functions to be compatible with one another in triple overlap regions. Thus if

$$x^a[p] = x^a(y^b[p]) \text{ in } U_{12} \tag{4.1.1}$$

$$y^a[p] = y^a(z^b[p]) \text{ in } U_{23} \tag{4.1.2}$$

$$x^a[p] = x^a(z^b[p]) \text{ in } U_{13} \tag{4.1.3}$$

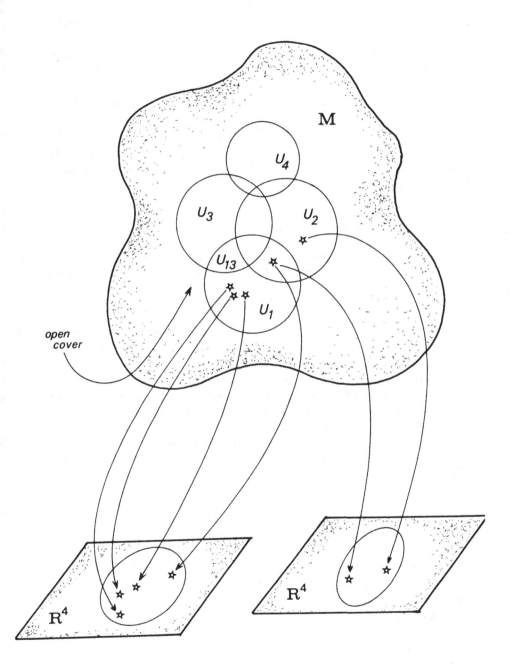

Figure 4.1. Covering of a four-dimensional manifold M by coordinate patches.

then we require that $x^a(y^b(z^c[p])) = x^a(z^c[p])$ holds in the triple overlap region $U_{123} = U_1 \cap U_2 \cap U_3$.

For a *differentiable manifold* we require the transition functions $x^a(y^b)$, etc., to be *differentiable* functions. In particular, if the transition functions have continuous partial derivatives of order r then we say the manifold is differentiable of class r. If the transition functions admit derivatives to all orders then we may say that the manifold is *smooth*.

Apart from the obvious example of $\mathbf{R}^n$ itself, some elementary examples of n-dimensional manifolds include the n-sphere S^n, the n-torus T^n (defined by the topological product of n circles), and real projective space RP^n (defined as the space of lines through the origin in R^{n+1}).

4.2   Scalar fields

Let $\phi(x^a)$ be a scalar field on U_1 and $\psi(y^a)$ a scalar field in U_2. If they agree on U_{12} then together they define a scalar field on the union $U_1 \cup U_2$. By 'agree' we mean:

$$\phi(x^a[p]) = \psi(y^a[p]) \tag{4.2.1}$$

for all p in U_{12}. Thus if we have a collection of scalar fields, one for each open set in M, such that they all agree as above in the overlap regions, then we obtain thereby a globally well-defined scalar field on the manifold.

In what follows we shall compactify the notation slightly by labelling a 'typical' pair of patches U and U' respectively with coordinates x^a and x'^a. Then if $\phi(x^a)$ is a scalar field on U and $\phi'(x'^a)$ is a scalar field on U', these agree if

$$\phi(x^a[p]) = \phi'(x'^a[p]) \text{ on } U \cap U',$$

or, equivalently, if

$$\phi(x^a) = \phi'(x'^b(x^a)).$$

4.3   Contravariant vector fields

These are defined in the analysis of differentiable manifolds through the consideration of linear differential operators acting on scalar fields. Intuitively, it is the 'direction' associated with the derivative at each point that thereby defines the directionality of the associated vector field.

Consider a pair of differential operators (acting on scalar fields) defined on sets U and U', given by

$$V^a(x)\frac{\partial}{\partial x^a} \text{ on } U,$$

$$V'^a(x')\frac{\partial}{\partial x'^a} \text{ on } U'.$$

When do these agree? Let $\phi(x)$ be a scalar field on U and $\phi'(x')$ a scalar field on U' such that $\phi(x) = \phi'(x')$ on $U \cap U'$. Then for any such scalar field we require

$$V^a(x)\frac{\partial}{\partial x^a}\phi(x) = V'^a(x')\frac{\partial}{\partial x'^a}\phi'(x')$$

for all p in $U \cap U'$; but

$$V^a(x)\frac{\partial}{\partial x^a}\phi(x)$$

$$= V^a(x)\frac{\partial x'^b}{\partial x^a}\frac{\partial}{\partial x'^b}\phi(x(x')) \quad \text{(chain rule)}$$

$$= V^a(x)\frac{\partial x'^b}{\partial x^a}\frac{\partial}{\partial x'^b}\phi'(x')$$

$$= V^b(x)\frac{\partial x'^a}{\partial x^b}\frac{\partial}{\partial x'^a}\phi'(x') \quad \text{(index rearrangement)}.$$

Therefore we require:

$$V'^a(x')\frac{\partial}{\partial x'^a}\phi'(x') = V^b(x)\frac{\partial x'^a}{\partial x^b}\frac{\partial}{\partial x'^a}\phi'(x') \text{ in } U \cap U',$$

for all $\phi'(x')$. Equating the coefficients of $\partial\phi'(x')/\partial x'^a$ we get

$$V'^a(x') = V^b(x)\frac{\partial x'^a}{\partial x^b} \text{ in } U \cap U'.$$

This is the *transition law* (or 'coordinate transformation law') for a *contravariant vector field*. It stipulates the transformation in the components of a contravariant vector field in the transition between two overlapping coordinate patches.

To define a contravariant vector field on a more extented portion of the manifold (or on the entirety of the manifold) we require an analogous transition relation on each of the relevant overlap regions.

For reasons that should be evident shortly we sometimes refer to a contravariant vector field as a *tensor field of valence* $[^1_0]$.

4.4   Covariant vector fields

These form a distinct system of vector fields on the manifold.   They are naturally 'dual' (in a sense to be described below) to contravariant vector fields. The transition law for covariant vector fields is as follows:

$$A'_a(x') = A_b(x)\frac{\partial x^b}{\partial x'^a} \text{ in } U \cap U'.$$

Note in particular that the Jacobian matrix $\partial x^b/\partial x'^a$ appears *inversely* relative to its appearance in the transition formula for contravariant vectors.

How do such fields arise? We consider an example. Let $\phi(x)$ be a scalar field with $\phi(x) = \phi'(x')$ in $U \cap U'$ for a typical overlap region. Consider its gradient, taken patch by patch, and set

$$A_a(x) = \frac{\partial\phi(x)}{\partial x^a}, \quad A'_a(x') = \frac{\partial\phi'(x')}{\partial x'^a},$$

and so on. How does this system transform? We have:

$$A_b(x) = \frac{\partial \phi(x)}{\partial x^b}$$

$$= \frac{\partial x'^c}{\partial x^b} \frac{\partial \phi(x(x'))}{\partial x'^c} \text{ in } U \cap U'$$

$$= \frac{\partial x'^c}{\partial x^b} \frac{\partial \phi'(x')}{\partial x'^c}$$

$$= \frac{\partial x'^c}{\partial x^b} A'_c(x').$$

Thus if we multiply each side by $\partial x^b / \partial x'^a$ and use the identity

$$\frac{\partial x^b}{\partial x'^a} \frac{\partial x'^c}{\partial x^b} = \delta_a^c$$

we obtain

$$A_b(x) \frac{\partial x^b}{\partial x'^a} = A'_a(x'),$$

as desired. We see therefore that the gradient of a scalar transforms as a *covariant* vector field.

A covariant vector field of course need not *necessarily* be a gradient—however, in n dimensions a covariant vector field can always locally be expressed in the form $A_a = \sum_r \Theta_r \nabla_a \phi_r$ where Θ_r and ϕ_r are scalars, and $r = 1 \ldots n$. In this case the scalars ϕ_r are chosen such that at each point their derivatives $\nabla_a \phi_r$ are linearly independent. Then the n linearly independent vectors $\nabla_a \phi_r$ constitute a *basis* for covector fields over the relevant region of the manifold (henceforth we shall use the helpful term *covector* as an abbreviation of 'covariant vector'), and the scalar fields Θ_r are the corresponding coefficients in the basis expansion for A_a.

In fact, there is a rather stronger result for the expression of covector fields in terms of gradients, known as *Pfaff's theorem*. The precise statement of the theorem varies according to whether the dimension of the manifold is even or odd. For n even let s run over the range $s = 1 \ldots \frac{1}{2}n$. Then for any covector field A_a there exists a set of scalars Θ_s and ϕ_s such that $A_a = \sum_s \Theta_s \nabla_a \phi_s$. So we only need half as many scalars as suggested in the previous paragraph! In odd dimensions we set $s = 1 \ldots \frac{1}{2}(n-1)$. Then for any covector A_a we have Θ_s, ϕ_s, and a scalar ψ such that $A_a = \nabla_a \psi + \sum_s \Theta_s \nabla_a \phi_s$. Thus again only n scalars are required.

4.5   Tensors in general

For general tensor fields we require multi-index quantities that transform according to appropriate transition laws in the coordinate patch overlap regions. A tensor field showing p contravariant indices and q covariant indices is said to have *valence* $\begin{bmatrix} p \\ q \end{bmatrix}$. In the expression of the transition law we have for each index either a Jacobian matrix, or the corresponding inverse matrix, depending on whether the index is of contra-variant or covariant type. The *total* valence of a tensor is called its *rank*.

We list below as examples the transition laws for several types of tensor fields:

$$\begin{bmatrix}3\\0\end{bmatrix} \quad A'^{abc}(x') = \frac{\partial x'^a}{\partial x^p}\frac{\partial x'^b}{\partial x^q}\frac{\partial x'^c}{\partial x^r}A^{pqr}(x)$$

$$\begin{bmatrix}0\\2\end{bmatrix} \quad B'_{ab}(x') = \frac{\partial x^p}{\partial x'^a}\frac{\partial x^q}{\partial x'^b}B_{pq}(x)$$

$$\begin{bmatrix}1\\1\end{bmatrix} \quad C'^{a}_{b}(x') = \frac{\partial x'^a}{\partial x^p}\frac{\partial x^q}{\partial x'^b}C^{p}_{q}(x).$$

As mentioned in §4.4, vector fields and covector fields are naturally dual. This means that there exists a well-defined scalar product between a vector field of each type. The *scalar product* between a vector field A^a and a covector field B_a is $A^a B_a$. Indeed it is straightforward to verify that the scalar product does transform like a scalar in overlap regions—we have

$$A'^a(x')B'_a(x') = A^a(x)\frac{\partial x'^b}{\partial x^a}\frac{\partial x^c}{\partial x'^b}B_c(x) = A^a(x)B_a(x),$$

by use of the identity

$$\frac{\partial x'^b}{\partial x^a}\frac{\partial x^c}{\partial x'^b} = \delta^c_a.$$

It should be evident that tensor fields can be multiplied together to produce fields of higher valence. For example, an equation in the form $P^{ab}_c = Q^a_c R^b$ is consistent with the transition laws if P^{ab}_c, Q^a_c and R^b are tensors of valence $\begin{bmatrix}2\\1\end{bmatrix}$, $\begin{bmatrix}1\\1\end{bmatrix}$, and $\begin{bmatrix}1\\0\end{bmatrix}$, respectively. Likewise, analogous to the formation of scalar product, we may *contract* tensors together in appropriate circumstances to produce tensors of lower rank. Thus the equation $S_{ab} = T_{abc}U^c$ is consistent with the transition laws if S_{ab}, T_{abc}, and U^c are tensors of valence $\begin{bmatrix}0\\2\end{bmatrix}$, $\begin{bmatrix}0\\3\end{bmatrix}$ and $\begin{bmatrix}1\\0\end{bmatrix}$ respectively. In summary, the basic operations of tensor algebra are:

(1) *addition*, e.g. $A^p_{qr} + B^p_{qr} = C^p_{qr}$

(2) *outer multiplication*, e.g. $A^p_q B^r_{st} = C^{pr}_{qst}$

(3) *inner multiplication (contraction)*, e.g. $L^m_n M^n_{op} = N^m_{op}.$

Exercises for chapter 4

[4.1] Show that if x^a and x'^a are coordinates for a manifold M (assumed differentiable) associated with the patches U and U', respectively, then in the overlap region $U \cap U'$ we have:

$$\frac{\partial x'^a}{\partial x^p}\frac{\partial^2 x^p}{\partial x'^b \partial x'^c} = -\frac{\partial x^q}{\partial x'^b}\frac{\partial x^r}{\partial x'^c}\frac{\partial^2 x'^a}{\partial x^q \partial x^r}.$$

[4.2] Suppose $A_a(x)$ transforms as a covariant vector field. Show that $\partial_{[a} A_{b]}$ transforms as a tensor of valence $\left[\begin{smallmatrix}0\\2\end{smallmatrix}\right]$, where $\partial_a = \partial/\partial x^a$. Similarly, show that if B_{ab} is a tensor field satisfying $B_{ab} = -B_{ba}$, then $\partial_{[a} B_{bc]}$ is also a tensor field.

[4.3] Suppose $P^a(x)$ and $Q^a(x)$ are contravariant vector fields. Show that

$$R^b = P^a \partial_a Q^b - Q^a \partial_a P^b$$

is also a contravariant vector field where $\partial_a = \partial/\partial x^a$.

[4.4] Suppose $P^a(x)$ is a contravariant vector field and S_a is a covariant vector field. Show that

$$T_b = P^a \partial_a S_b + S_a \partial_b P^a$$

is also a covariant field, where $\partial_a = \partial/\partial x^a$.

Remarks: the operation noted in exercise [4.2] is called the *exterior derivative* (cf. chapter 11); the operations of exercises [4.3] and [4.4] are examples of the so called *Lie derivative* (cf. chapter 8). Both of these derivation operations are purely natural or 'intrinsic' to the structure of a differentiable manifold. More general differentiation formulae, however, require the addition of more structure (a *connection*) as described in the next chapter.

5   Covariant differentiation

5.1   Making a connection

IT IS EVIDENT from the foregoing discussion that differentiation is a well-defined operation on manifolds, provided that the transition functions for the manifold are themselves differentiable. However, the differentiation operation thereby defined consists simply of a map from scalar functions to covector fields. Is there some procedure by which the scope of differentiation can be enlarged so as to include acts of differentiation defined on tensor fields? If so, we might imagine that such an operation would consistently map tensor fields of valence $\begin{bmatrix} p \\ q \end{bmatrix}$ to fields of valence $\begin{bmatrix} p \\ q+1 \end{bmatrix}$.

This can be done—but only at the expense of the introduction of extra structure on the manifold. The extra structure we require is called a *connection*, and a manifold thus endowed is called a *manifold with connection*. Once a manifold is endowed with a connection, differentiation takes on a well-defined meaning for tensor fields of any valence.

Suppose M is covered by a family of open sets U, U', and so on. On each patch we define a three-index array of functions which we shall denote $\Gamma^a_{bc}(x)$ on U, $\Gamma'^a_{bc}(x')$ on U', and so on.

In transition regions these quantities are *not* required to transform like tensors, but rather to satisfy the following *modified* transition relations:

$$\Gamma'^a_{bc} = \frac{\partial x'^a}{\partial x^p} \frac{\partial x^q}{\partial x'^b} \frac{\partial x^r}{\partial x'^c} \Gamma^p_{qr} + \frac{\partial x'^a}{\partial x^p} \frac{\partial^2 x^p}{\partial x'^b \partial x'^c}$$

for a typical transition region $U \cap U'$. The significance of the apparently complicated transition rule will be made clear shortly. Note that by virtue of

$$\frac{\partial x'^b}{\partial x^a} \frac{\partial x^c}{\partial x'^b} = \delta^c_a$$

the second term on the right can be rewritten as

$$\frac{\partial x'^a}{\partial x^p} \frac{\partial^2 x^p}{\partial x'^b \partial x'^c} = -\frac{\partial x^q}{\partial x'^b} \frac{\partial x^r}{\partial x'^c} \frac{\partial^2 x'^a}{\partial x^q \partial x^r}.$$

Thus if we write $J^a_b = \partial x'^a / \partial x^b$ and $\tilde{J}^a_b = \partial x^a / \partial x'^b$ we have $J^a_b \tilde{J}^b_c = \delta^a_c$, and for our transition rules we may write equivalently either

$$\Gamma'^a_{bc} = J^a_p \tilde{J}^q_b \tilde{J}^r_c \Gamma^p_{qr} + J^a_p \partial'_b \tilde{J}^p_c \qquad (5.1.1)$$

or (cf. exercise 4.4):

$$\Gamma_{bc}^{\prime a} = J_p^a \tilde{J}_b^q \tilde{J}_c^r \Gamma_{qr}^p - \tilde{J}_b^q \tilde{J}_c^r \partial_q J_r^a. \tag{5.1.2}$$

Here we employ the useful abbreviations $\partial_a = \partial/\partial x^a$ and $\partial'_a = \partial/\partial x^{\prime a}$ which have the virtue of presenting the relevant indices in an obvious way in the correct position.

Once such a field has been specified on M we have a *manifold with connection*. But what about triple overlap regions? Are there not compatibility relations to be satisfied amongst the various Γ_{bc}^a quantities in such regions?

Remarkably—and this is what accounts for the specific form of the 'inhomogeneous' term appearing in equation (5.1.1)—the compatibility conditions are *automatically satisfied*.

Indeed, suppose we have three patches U, U', U'' , with associated coordinates x^a, $x^{\prime a}$, $x^{\prime\prime a}$. In the triple overlap region $U \cap U' \cap U''$ suppose Γ_{bc}^a transforms to $\Gamma_{bc}^{\prime a}$ and $\Gamma_{bc}^{\prime a}$ transforms to $\Gamma_{bc}^{\prime\prime a}$. Then it turns out that Γ_{bc}^a automatically transforms correctly to $\Gamma_{bc}^{\prime\prime a}$. To see this we proceed as follows. Let us write

$$
\begin{array}{lll}
J_b^a = \partial x^{\prime a}/\partial x^b & \tilde{J}_b^a = \partial x^a/\partial x^{\prime b} & J_b^a \tilde{J}_c^b = \delta_c^a \\
K_b^a = \partial x^{\prime\prime a}/\partial x^{\prime b} & \tilde{K}_b^a = \partial x^{\prime a}/\partial x^{\prime\prime b} & K_b^a \tilde{K}_c^b = \delta_c^a \\
L_b^a = \partial x^{\prime\prime a}/\partial x^b & \tilde{L}_b^a = \partial x^a/\partial x^{\prime\prime b} & L_b^a \tilde{L}_c^b = \delta_c^a.
\end{array}
$$

Thus given (5.1.1) together with

$$\Gamma_{bc}^{\prime\prime a} = K_p^a \tilde{K}_b^q \tilde{K}_c^r \Gamma_{qr}^{\prime p} + K_p^a \partial_b^{\prime\prime} \tilde{K}_c^p \tag{5.1.3}$$

where $\partial_a^{\prime\prime} = \partial/\partial x^{\prime\prime a}$, we wish to prove that:

$$\Gamma_{bc}^{\prime\prime a} = L_u^a \tilde{L}_b^v \tilde{L}_c^w \Gamma_{vw}^u + L_u^a \partial_b^{\prime\prime} \tilde{L}_c^u \tag{5.1.4}$$

in the triple overlap region. Suppose the expression $\Gamma_{qr}^{\prime p}$ in (5.1.1) is substituted into (5.1.3). We obtain:

$$
\begin{aligned}
\Gamma_{bc}^{\prime\prime a} &= K_p^a \tilde{K}_b^q \tilde{K}_c^r (J_u^p \tilde{J}_q^v \tilde{J}_r^w \Gamma_{vw}^u + J_u^p \partial_q' \tilde{J}_r^u) + K_p^a \partial_b^{\prime\prime} \tilde{K}_c^p \\
&= L_u^a \tilde{L}_b^v \tilde{L}_c^w \Gamma_{vw}^u + K_p^a \tilde{K}_b^q \tilde{K}_c^r J_u^p \partial_q' \tilde{J}_r^u + K_p^a \partial_b^{\prime\prime} \tilde{K}_c^p \\
&= L_u^a \tilde{L}_b^v \tilde{L}_c^w \Gamma_{vw}^u + L_u^a (\tilde{K}_c^r \partial_b^{\prime\prime} \tilde{J}_r^u) + K_q^a J_u^a (\tilde{J}_r^u \partial_b^{\prime\prime} \tilde{K}_c^r) \\
&= L_u^a \tilde{L}_b^v \tilde{L}_c^w \Gamma_{vw}^u + L_u^a \partial_b^{\prime\prime} (\tilde{K}_c^r \tilde{J}_r^u) \\
&= L_u^a \tilde{L}_b^v \tilde{L}_c^w \Gamma_{vw}^u + L_u^a \partial_b^{\prime\prime} \tilde{L}_c^u
\end{aligned}
$$

as desired, where use has been made of the relations $K_b^a J_c^b = L_c^a$, and $\tilde{K}_b^q \partial_q' = (\partial_b^{\prime\prime} x^{\prime q}) \partial_q = \partial_b^{\prime\prime}$. So we see that our transition relations for the connection are compatible in triple overlap regions.

At first glance this calculation may give the impression of being rather difficult—but the complexities are only superficial, and one should be impressed, rather, with the power and efficiency of the operations involved.

In summary, a *manifold with connection* is a differentiable manifold endowed with a field Γ_{bc}^a subject to transition laws of the form (5.1.1). Note that Γ_{bc}^a is not itself a tensor field (the presence of the inhomogeneous term in the transition law prevents this). Nevertheless it is precisely what is needed in order to form tensor fields from tensor fields by differentiation.

5.2 Differentiation of tensors.

The specification of a connection is in fact equivalent to the stipulation of a *general rule for the differentiation of tensor fields*. To see this note first that if V^a is a contravariant vector field then the elementary expression $\partial_b V^a$ *fails* to transform as a tensor field of valence $\begin{bmatrix}1\\1\end{bmatrix}$ in transition regions. One obtains:

$$\partial'_b V'^a = \tilde{J}^p_b \partial_p (J^a_q V^q) = \tilde{J}^p_b J^a_q (\partial_p V^q) + (\tilde{J}^p_b \partial_p J^a_q) V^q$$

with an extra term appearing that contradicts the transition law for a tensor field of valence $\begin{bmatrix}1\\1\end{bmatrix}$.

Consider now the following expression, which defines the action of *covariant differentiation* on vector fields of valence $\begin{bmatrix}1\\0\end{bmatrix}$:

$$\nabla_b V^a = \partial_b V^a + \Gamma^a_{bc} V^c. \tag{5.2.1}$$

In a typical transition region $U \cap U'$ we find that $\nabla'_b V'^a = \tilde{J}^q_b J^a_p (\nabla_q V^p)$, as desired, as follows from

$$
\begin{aligned}
\nabla'_b V'^a &= \partial'_b V'^a + \Gamma'^a_{bc} V'^c \\
&= \tilde{J}^q_b J^a_p (\partial_q V^p) + (\tilde{J}^q_b \partial_q J^a_p) V^p + (J^a_p \tilde{J}^q_b \tilde{J}^r_c \Gamma^p_{qr} - \tilde{J}^q_b \tilde{J}^r_c \partial_q J^a_r) J^c_s V^s \\
&= \tilde{J}^q_b J^a_p (\partial_q V^p) + \tilde{J}^q_b J^a_p (\tilde{J}^r_c J^c_s \Gamma^p_{qr} V^s) + (\tilde{J}^q_b \partial_q J^a_p) V^p - (\tilde{J}^q_b \partial_q J^a_r)(\tilde{J}^r_c J^c_s V^s) \\
&= \tilde{J}^q_b J^a_p (\partial_q V^p) + \tilde{J}^q_b J^a_p (\Gamma^p_{qr} V^r) \\
&= \tilde{J}^q_b J^a_p (\nabla_q V^p)
\end{aligned}
$$

by use of (5.1.2) and $\tilde{J}^a_b J^b_c = \delta^a_c$. Thus we see that if V^a transforms as a tensor of valence $\begin{bmatrix}1\\0\end{bmatrix}$, then $\nabla_b V^a$ transforms as a tensor of valence $\begin{bmatrix}1\\1\end{bmatrix}$. The inhomogeneous term in (5.1.1) is just what is required to cancel the inhomogeneous term in the partial derivative $\partial'_b V'^a$. For *covector fields* the appropriate formula for differentiation is

$$\nabla_b W_a = \partial_b W_a - \Gamma^c_{ba} W_c. \tag{5.2.2}$$

Again we may verify (as above) that $\nabla_b W_a$ is a tensor of valence $\begin{bmatrix}0\\2\end{bmatrix}$. The formula given here for $\nabla_b W_a$ is motivated in part by the requirement that ∇_b should satisfy a *Leibniz* property when it acts on products of vectors and covectors. In particular, for scalar fields we define $\nabla_a \phi = \partial_a \phi$. This is consistent with the Leibniz property for differentiation of a product. For if $\phi = V^b W_b$ we have

$$
\begin{aligned}
\nabla_a \phi &= \nabla_a (V^b W_b) \\
&= (\nabla_a V^b) W_b + V^b (\nabla_a W_b) \quad \text{[Leibniz]} \\
&= (\partial_a V^b + \Gamma^b_{ac} V^c) W_b + V^b (\partial_a W_b - \Gamma^c_{ab} W_c) \\
&= (\partial_a V^b) W_b + V^b \partial_a W_b = \partial_a (V^b W_b),
\end{aligned}
$$

which agrees with $\nabla_a \phi = \partial_a \phi$. The general formula for covariant differentiation is given, e.g., according to the scheme:

$$\nabla_a V^{bc}_d = \partial_a V^{bc}_d + \Gamma^b_{ap} V^{pc}_d + \Gamma^c_{ap} V^{bp}_d - \Gamma^q_{ad} V^{bc}_q \tag{5.2.3}$$

the corresponding formulae for other valences being constructed analogously.

5.3   Torsion

Suppose ϕ is a smooth scalar field. Then clearly in each patch we have $\partial_a \partial_b \phi = \partial_b \partial_a \phi$. It is important to note however that $\nabla_a \nabla_b \phi$ does *not* necessarily equal $\nabla_b \nabla_a \phi$. One can show, in fact, that there always exists a tensor T_{ab}^c (the *torsion* tensor) such that for any smooth scalar field we have:

$$(\nabla_a \nabla_b - \nabla_b \nabla_a)\phi = T_{ab}^c \nabla_c \phi. \qquad (5.3.1)$$

If $T_{ab}^c = 0$ then the connection is said to be *torsion-free*. In Einstein's theory of gravitation it will be an *assumption* that the torsion vanishes; but one can consider more general theories of gravity, e.g. the so-called *Einstein-Cartan theory* where T_{ab}^c is not assumed to vanish and is given physical interpretation. Let us now calculate T_{ab}^c in terms of Γ_{ab}^c. Recall that $\nabla_a V_b = \partial_a V_b - \Gamma_{ab}^c V_c$ for any covariant vector field V_b. Consequently if we set $V_b = \nabla_b \phi = \partial_b \phi$ we get

$$\nabla_a \nabla_b \phi = \partial_a \partial_b \phi - \Gamma_{ba}^c \partial_c \phi, \qquad \nabla_b \nabla_a \phi = \partial_b \partial_a \phi - \Gamma_{ab}^c \partial_c \phi.$$

Whence by subtraction, we obtain: $(\nabla_a \nabla_b - \nabla_b \nabla_a)\phi = T_{ab}^c \nabla_c \phi$ with

$$T_{ab}^c = -\Gamma_{ab}^c + \Gamma_{ba}^c = -2\Gamma_{[ab]}^c. \qquad (5.3.2)$$

From (5.1) we see that although Γ_{ab}^c is not a tensor nevertheless $\Gamma_{[bc]}^a$ *is* a tensor. Thus the torsion is a measure of the *skew-symmetric* part of the connection. The torsion-free condition therefore amounts to an assumption that the connection is *symmetric*.

Exercises for chapter 5

[5.1] If $W_a(x)$ is a tensor field of valence $\begin{bmatrix} 0 \\ 1 \end{bmatrix}$ verify by direct calculation that $\nabla_b W_a$ is a tensor field of valence $\begin{bmatrix} 0 \\ 2 \end{bmatrix}$, showing that it has the correct transformational properties between patches.

[5.2] A manifold M is endowed with a connection such that

$$(\nabla_a \nabla_b - \nabla_b \nabla_a)V_c = T_{ab}^d \nabla_d V_c$$

for any smooth vector field V_a. Show that under these conditions T_{ab}^c satisfies

$$\nabla_{[a} T_{bc]}^d + T_{[ab}^r T_{c]r}^d = 0.$$

6   Properties of Riemann tensor

6.1   What is the Riemann tensor?

WE ARE NOW ready to introduce a fundamental tensor of valence $\begin{bmatrix} 1 \\ 3 \end{bmatrix}$ that exists for any manifold with connection. Our only stipulation is that the connection should be at least once differentiable—i.e. that $\partial_a \Gamma^b_{cd}$ should exist in each patch. In Einstein's theory this tensor can be thought of as representing the *gravitational field*: it plays a role in many respects parallel to that played by the electromagnetic field strength F_{ab} in Maxwell's theory.

The Riemann tensor arises as a measure of the extent to which commutativity fails for the second covariant derivative of a vector field. The definition is best given implicitly in the form of the following theorem:

(6.1.1) **Theorem.** *Let M be a manifold with a connection Γ^c_{ab} that is at least once differentiable in each coordinate patch. Then there exists a unique tensor field $R_{abc}{}^d$ such that:*

$$(\nabla_a \nabla_b - \nabla_b \nabla_a) V_c - T^d_{ab} \nabla_d V_c = -R_{abc}{}^d V_d$$

for any smooth vector field V_c, where T^d_{ab} is the torsion associated with ∇_a.

Proof: The calculation proceeds as follows. We wish to evaluate $\nabla_a \nabla_b V_c$ and $\nabla_b \nabla_a V_c$ for an arbitrary covector field V_c, and take the difference. We have:

$$\nabla_a \nabla_b V_c = \partial_a(\nabla_b V_c) - \Gamma^p_{ab} \nabla_p V_c - \Gamma^p_{ac} \nabla_b V_p$$
$$\nabla_b \nabla_a V_c = \partial_b(\nabla_a V_c) - \Gamma^p_{ba} \nabla_p V_c - \Gamma^p_{bc} \nabla_a V_p$$

whence

$$(\nabla_a \nabla_b - \nabla_b \nabla_a) V_c + (\Gamma^p_{ab} \nabla_p V_c - \Gamma^p_{ba} \nabla_p V_c)$$
$$= \partial_a(\nabla_b V_c) - \partial_b(\nabla_a V_c) - \Gamma^p_{ac} \nabla_b V_p + \Gamma^p_{bc} \nabla_a V_p.$$

Since the torsion tensor is defined by $T^c_{ab} = -\Gamma^c_{ab} + \Gamma^c_{ba}$ we obtain:

$$(\nabla_a \nabla_b - \nabla_b \nabla_a) V_c - T^d_{ab} \nabla_d V_c$$
$$= \partial_a(\nabla_b V_c) - \partial_b(\nabla_a V_c) - \Gamma^p_{ac} \nabla_b V_p + \Gamma^p_{bc} \nabla_a V_p.$$

The four terms on the right can be evaluated as follows:

$$\partial_a(\nabla_b V_c) = \partial_a \partial_b V_c - \partial_a(\Gamma^d_{bc} V_d)$$

$$= \partial_a \partial_b V_c - (\partial_a \Gamma^d_{bc}) V_d - \Gamma^d_{bc}(\partial_a V_d), \tag{6.1.2}$$

$$\partial_b(\nabla_a V_c) = \partial_b \partial_a V_c - (\partial_b \Gamma^d_{ac}) V_d - \Gamma^d_{ac}(\partial_b V_d), \tag{6.1.3}$$

$$\Gamma^p_{ac} \nabla_b V_p = \Gamma^p_{ac} \partial_b V_p - \Gamma^p_{ac} \Gamma^d_{bp} V_d, \tag{6.1.4}$$

$$\Gamma^p_{bc} \nabla_a V_p = \Gamma^p_{bc} \partial_a V_p - \Gamma^p_{bc} \Gamma^d_{ap} V_d. \tag{6.1.5}$$

Combining these in the order given, i.e. $(6.1.2) - (6.1.3) - (6.1.4) + (6.1.5)$, leads to the result of the theorem, with

$$\frac{1}{2} R_{abc}{}^d = \partial_{[a} \Gamma^d_{b]c} - \Gamma^p_{[a|c|} \Gamma^d_{b]p} \tag{6.1.6}$$

where in the final term we employ the following alternative notation for skew-symmetrization:

$$A_{[a|b|c]d} = \frac{1}{2}(A_{abcd} - A_{cbad})$$

which is convenient when complicated expressions are involved.

We see that although Γ^c_{ab} is not a tensor, nevertheless the particular combination of derivatives and non-linear terms appearing in (6.1.6) is such that $R_{abc}{}^d$ does transform appropriately in overlap regions: it is a tensor of the greatest significance.

6.2   The Ricci identities

These are the various important identities that define the Riemann tensor in terms of the skew-symmetrized second covariant derivative of an arbitrary vector field. From the discussion above we already have

$$\nabla_{[a} \nabla_{b]} V_c = \frac{1}{2} T^d_{ab} \nabla_d V_c - \frac{1}{2} R_{abc}{}^d V_d \tag{6.2.1}$$

for *covariant* vectors. For *contravariant* vectors we have

$$\nabla_{[a} \nabla_{b]} W^d = \frac{1}{2} T^c_{ab} \nabla_c W^d + \frac{1}{2} R_{abc}{}^d W^c. \tag{6.2.2}$$

Note carefully the *change of sign* here in the coefficient of the Riemann tensor. The result for contravariant vectors follows at once from the result for covariant vectors by use of

$$\nabla_{[a} \nabla_{b]} \phi = \frac{1}{2} T^c_{ab} \nabla_c \phi \tag{6.2.3}$$

with $\phi = V_c W^c$. The action of $\nabla_{[a} \nabla_{b]}$ on tensors of higher valence is exemplified as follows:

$$\nabla_{[a} \nabla_{b]} U^{de}_c = \frac{1}{2} T^f_{ab} \nabla_f U^{de}_c + \frac{1}{2} R_{abp}{}^d U^{pe}_c$$
$$+ \frac{1}{2} R_{abq}{}^e U^{dq}_c - \frac{1}{2} R_{abc}{}^r U^{de}_r. \tag{6.2.4}$$

Other examples can be readily constructed by this pattern.

6.3 Remarkable symmetries of the Riemann tensor

The symmetry $R_{abc}{}^d = -R_{bac}{}^d$ follows at once from the definition of the Riemann tensor. Another symmetry that follows in the *torsion-free* case is given below:

(6.3.1) **Proposition.** *If* ∇_a *is torsion-free then* $R_{[abc]}{}^d = 0$.

Proof. Since ∇_a is assumed torsion-free we have

$$\nabla_{[a}\nabla_{b]}V_c = -\frac{1}{2}R_{abc}{}^d V_d \qquad (6.3.2)$$

for any V_c, and if $V_c = \nabla_c \psi$ we obtain

$$\nabla_{[a}\nabla_b\nabla_{c]}\psi = -\frac{1}{2}R_{[abc]}{}^d\nabla_d\psi \qquad (6.3.3)$$

But $\nabla_{[a}\nabla_b\nabla_{c]}\psi = \nabla_{[a[}\nabla_b\nabla_{c]]}\psi = 0$ since ∇_a is torsion free. Therefore $R_{[abc]}{}^d\nabla_d\psi$ vanishes for *any* choice of ψ; whence $R_{[abc]}{}^d$ must vanish.

Note that $R_{[abc]}{}^d = \frac{1}{3}(R_{abc}{}^d + R_{bca}{}^d + R_{cab}{}^d)$, as follows from the fact that $R_{abc}{}^d$ is already antisymmetric over the first two of its indices. In exercise [6.4], proposition (6.3.1) is extended to include torsion. Cf. also exercise [5.2].

6.4 The Bianchi identity

We recall that an electromagnetic field tensor F_{ab} is taken to satisfy $\nabla_{[a}F_{bc]} = 0$ whether or not the vacuum condition applies. Remarkably the Riemann tensor *automatically* satisfies a similar identity.

(6.4.1) **Proposition.** *If* ∇_a *is torsion-free then* $\nabla_{[a}R_{bc]e}{}^d = 0$.

Proof. First we calculate, for arbitrary V^d, that

$$2\nabla_{[a}\nabla_{b]}\nabla_c V^d = -R_{abc}{}^p\nabla_p V^d + R_{abq}{}^d\nabla_c V^q$$

whence

$$2\nabla_{[a}\nabla_b\nabla_{c]}V^d = R_{[ab|q|}{}^d\nabla_{c]}V^q \qquad (6.4.2)$$

by $R_{[abc]}{}^d = 0$. Next we calculate

$$2\nabla_a\nabla_{[b}\nabla_{c]}V^d = (\nabla_a R_{bcq}{}^d)V^q + R_{bcq}{}^d\nabla_a V^q$$

whence

$$2\nabla_{[a}\nabla_b\nabla_{c]}V^d = \nabla_{[a}R_{bc]q}{}^d V^q + R_{[bc|q|}{}^d\nabla_{a]}V^q. \qquad (6.4.3)$$

Equating (6.4.2) and (6.4.3) we obtain the desired result. The relation just established is called the *Bianchi identity*:

$$\nabla_{[a}R_{bc]e}{}^d = 0.$$

It holds quite generally for any torsion-free connection, and is of great importance in general relativity theory. When torsion is included a modified identity can be established as in exercise [6.5].

Exercises for chapter 6

[6.1] If F_{ab} is a tensor field satisfying

$$F_{ab} = F_{[ab]} = \frac{1}{2}(F_{ab} - F_{ba})$$

show that for a torsion-free connection ∇_a we have

$$\nabla_{[a}\nabla_{b]}F_{cd} = \alpha R_{ab[c}{}^e F_{d]e}$$

where α is a constant that should be computed.

[6.2] If ∇_a and $\tilde{\nabla}_a$ are connections with

$$\tilde{\nabla}_a V^b = \nabla_a V^b + Q^b_{ac}V^c$$

for all vector fields V^b, show that

$$\frac{1}{2}(\tilde{R}_{abc}{}^d - R_{abc}{}^d) = \nabla_{[a}Q^d_{b]c} + Q^d_{r[a}Q^r_{b]c}$$

where $R_{abc}{}^d$ and $\tilde{R}_{abc}{}^d$ are the Riemann tensors associated with ∇_a and $\tilde{\nabla}_a$, respectively.

[6.3] The contracted Riemann tensor or *Ricci tensor* R_{ab} is defined by $R_{ab} = R_{apb}{}^p$. If ∇_a is torsion-free show that

$$\nabla_{[a}R_{b]c} = -\frac{1}{2}\nabla_d R_{abc}{}^d.$$

Show that R_{ab} is *not necessarily symmetric*. (N.B.—if ∇_a is the Levi-Civita connection described in chapter 7 then R_{ab} *is* symmetric.)

[6.4] Show that if the torsion does *not* vanish then:

$$R_{[abc]}{}^d + \nabla_{[a}T^d_{bc]} + T^p_{[ab}T^d_{c]p} = 0.$$

[6.5] Show that if the torsion does *not* vanish then:

$$\nabla_{[a}R_{bc]r}{}^s + T^d_{[ab}R_{c]dr}{}^s = 0.$$

[6.6] Suppose that M is an n-dimensional differentiable manifold with symmetric connection Γ^c_{ab}. Define:

$$P_{ij} = -\frac{n}{n^2-1}R_{ij} - \frac{1}{n^2-1}R_{ji}$$

where R_{ij} is the Ricci tensor (see exercise 6.3). Note that if R_{ij} is symmetric, then so is P_{ij}. M is said to be *projectively flat* if Γ^c_{ab} can be put in the form:

$$\Gamma^c_{ab} = \phi_a \delta^c_b + \phi_b \delta^c_a.$$

Show that if M is projectively flat then

(1) $R_{jki}{}^h + \delta^h_k P_{ji} - \delta^h_j P_{ki} - (P_{kj} - P_{jk})\delta^h_i = 0$

(2) $\nabla_k P_{ji} = \nabla_j P_{ki}$.

For $n > 2$ show that (2) is a consequence of (1). For $n = 2$ show that (1) is an identity. Show that if M is projectively flat then $P_{ij} = P_{ji}$ if $\phi_i = \nabla_i \phi$ for some scalar ϕ. Remark: (1) and (2) are also *sufficient* conditions for M to be projectively flat.

[6.7] Let $\Delta_{ab} = 2\nabla_{[a}\nabla_{b]}$ and assume the torsion vanishes, i.e. $\Delta_{ab}\phi = 0$, for any scalar ϕ. Show that the Leibniz rule for ∇_a implies a Leibniz rule for Δ_{ab}, i.e. show that:

$$\Delta_{ab}(PQ) = P\Delta_{ab}Q + Q\Delta_{ab}P$$

where P and Q (indices not shown) are tensors of any valence.

[6.8] Let $R_{abc}{}^d$ be the Riemann tensor associated with a connection Γ^c_{ab} (not necessarily symmetric). A new connection is defined by

$$\tilde{\Gamma}^c_{ab} = \Gamma^c_{ab} + \delta^c_a \partial_b \Phi.$$

Show that $\tilde{R}_{abc}{}^d = R_{abc}{}^d$ where $\tilde{R}_{abc}{}^d$ is the Riemann tensor associated with the new connection.

7   Riemannian geometry

7.1   The fundamental theorem of Riemannian geometry

IN RELATIVITY THEORY it is convenient to distinguish three levels or layers of geometry, with increasing structure and complexity: (a) the underlying differentiable manifold, (b) the connection, (c) the metric. By the latter we mean a second-rank symmetric non-degenerate tensor field g_{ab}, which therefore defines an inner-product operation on pairs of contravariant vector fields.

For *Riemannian* geometry (b) and (c) are made compatible, in a special sense, by the requirement $\nabla_a g_{bc} = 0$, with ∇_a torsion-free. Thus for a Riemannian geometry the metric tensor is *covariantly constant* with respect to the connection ∇_a.

Remarkably, these conditions are sufficient to determine ∇_a uniquely in terms of g_{ab}, and it is this natural link between metric and connection that constitutes the foundation of Riemannian geometry.

(7.1.1) **Theorem.** *Let g_{ab} be the metric, a symmetric, non-degenerate, differentiable tensor field, on a manifold M. There exists a unique torsion-free connection ∇_a on M such that $\nabla_a g_{bc} = 0$.*

Proof: In order for $\nabla_a g_{bc}$ to vanish we must have

$$\partial_a g_{bc} - \Gamma^{b'}_{ab} g_{b'c} - \Gamma^{c'}_{ac} g_{bc'} = 0. \tag{7.1.2}$$

Note that c' is of course a distinct 'letter' from c, but the prime helps to remind us which letter has been 'displaced' in the operation of covariant differentiation. By cyclic permutation of indices we also have:

$$\partial_b g_{ca} - \Gamma^{c'}_{bc} g_{c'a} - \Gamma^{a'}_{ba} g_{ca'} = 0 \tag{7.1.3}$$

and

$$\partial_c g_{ab} - \Gamma^{a'}_{ca} g_{a'b} - \Gamma^{b'}_{cb} g_{ab'} = 0. \tag{7.1.4}$$

If from these equations we form the combination (7.1.2) minus (7.1.3) minus (7.1.4) various terms cancel out, and we get

$$2\Gamma^{c'}_{bc} g_{c'a} + \partial_a g_{cb} - \partial_b g_{ca} - \partial_c g_{ab} = 0 \tag{7.1.5}$$

after use of the torsion-free condition $\Gamma^a_{bc} = \Gamma^a_{cb}$.

Since g_{ab} is non-degenerate there exists a unique symmetric tensor g^{ab} such that $g_{ab}g^{bc} = \delta^c_a$. Contracting (7.1.5) with g^{ad} we get:

$$\Gamma^d_{bc} = \frac{1}{2}g^{da}(\partial_b g_{ca} + \partial_c g_{ba} - \partial_a g_{bc}). \tag{7.1.6}$$

Thus if $\nabla_a g_{bc} = 0$ then Γ^c_{ab} must certainly be of this form. Conversely if Γ^d_{bc} is given as above, then one can show by direct calculation that $\nabla_a g_{bc} = 0$, as desired.

The connection thus constructed is variously called the 'metric connection', the 'Levi-Civita connection', or the 'Riemannian connection'.

7.2   Symmetries of the curvature tensor

Let ∇_a be the torsion-free Levi-Civita connection associated with g_{ab} (as will be assumed henceforth, unless it is otherwise stated), and let $R_{abc}{}^d$ be the Riemann tensor associated in the usual way with ∇_a, as described in chapter 6. The metric tensors g^{ab} and g_{ab} may be used to raise and lower indices. Moreover, since $\nabla_a g_{bc} = 0$ the process of raising and lowering indices naturally 'commutes' with differentiation. Thus it is natural to lower the final index of the Riemann tensor $R_{abc}{}^d$ and consider properties of the tensor R_{abcd} defined by

$$R_{abcd} = R_{abc}{}^{d'} g_{dd'}.$$

We shall refer to the Riemann tensor associated with a metric g_{ab} as its *curvature tensor*. Here we shall record a number of special symmetries possessed by the curvature tensor;—their mastery is essential to a practical understanding of gravitation. These symmetries are important inasmuch as they make frequent appearance in algebraic aspects of calculations involving the curvature tensor. The relevant symmetries may be summarized as follows:

$$R_{abcd} = R_{[ab]cd} \tag{7.2.1}$$

$$R_{[abc]d} = 0 \tag{7.2.2}$$

$$R_{abcd} = R_{ab[cd]} \tag{7.2.3}$$

$$R_{abcd} = R_{cdab}. \tag{7.2.4}$$

Now we already know (7.2.1) and (7.2.2) to be the case, and these properties are simply noted here again for convenience. For (7.2.3) we deduce from $\nabla_a g_{cd} = 0$ that $\nabla_{[a}\nabla_{b]}g_{cd} = 0$; but

$$2\nabla_{[a}\nabla_{b]}g_{cd} = -R_{abc}{}^{c'}g_{c'd} - R_{abd}{}^{d'}g_{d'c}$$

$$= -(R_{abcd} + R_{abdc})$$

whence $R_{abcd} + R_{abdc} = 0$, q.e.d.. To show (7.2.4) we observe that $R_{[abc]d} = 0$ implies

$$R_{abcd} = -R_{bcad} - R_{cabd}$$

$$= R_{cbad} - R_{cabd}$$

$$= -2R_{c[ab]d}$$

$$= -2R_{c|[ab]|d}. \tag{7.2.5}$$

Also

$$R_{abcd} = -R_{bcad} - R_{cabd}$$
$$= R_{bcda} - R_{acdb}$$
$$= -R_{acdb}$$
$$= -2R_{[a|[cd]|b]}$$

whence

$$R_{cdab} = -2R_{[c|[ab]|d]}. \tag{7.2.6}$$

The equation of (7.2.5) and (7.2.6) then yields the desired result.

7.3   The contracted Bianchi identity

Two more basic tensors can be defined in terms of the curvature tensor. These are

$$\text{The Ricci tensor:}\quad R_{ab} = R_{acb}{}^c = R_{acbd}g^{cd}, \tag{7.3.1}$$
$$\text{The Ricci scalar:}\quad R = R_a^a = R_{ab}g^{ab} = R_{acbd}g^{ab}g^{cd}. \tag{7.3.2}$$

Note that as a consequence of the identity (7.2.4) the Ricci tensor for a Riemannian geometry is necessarily symmetric: $R_{ab} = R_{ba}$. (Cf. however exercise 6.3.)

(7.3.3) **Proposition.** $\nabla^a(R_{ab} - \frac{1}{2}g_{ab}R) = 0.$

Proof. We begin with the Bianchi identity $\nabla_{[a}R_{bc]de} = 0$, that is

$$\nabla_a R_{bcde} + \nabla_b R_{cade} + \nabla_c R_{abde} = 0.$$

Contracting this relation with $g^{ad}g^{ce}$ we get

$$2\nabla^a R_{ab} - \nabla_b R = 0,$$

from which the desired result follows.

The *Einstein tensor* G_{ab} is defined by $G_{ab} = R_{ab} - \frac{1}{2}g_{ab}R$. It is therefore, by the contracted Bianchi identity, found to be *divergence-free:* $\nabla^a G_{ab} = 0$. This is a result of fundamental significance, and a primary basis for the arguments leading to Einstein's gravitational equations.

In Einstein's theory the *vacuum equations* (i.e. the gravitational field equations for regions of space-time absent of matter) are given by $R_{ab} = 0$. When matter is present we shall require G_{ab} to be *proportional* to T_{ab} (where T_{ab} is the stress-energy tensor for the matter configuration). Note that $R_{ab} = 0$ if and only if $G_{ab} = 0$.

Exercises for chapter 7

Note: Henceforth we assume (unless it is otherwise stated) that the *torsion vanishes*.

[7.1] Verify directly that if $\Gamma^d_{bc} = \frac{1}{2}g^{da}(\partial_b g_{ca} + \partial_c g_{ba} - \partial_a g_{bc})$ then $\nabla_a g_{bc} = 0$.

[7.2] Show that if $\nabla_a g_{bc} = 0$ then $\nabla_a g^{bc} = 0$.

[7.3] Show that if $R_{ab} = 0$ then the Riemann tensor satisfies $\nabla^a R_{abcd} = 0$.

[7.4] Suppose the vector field ξ_a satisfies Killing's equation: $\nabla_{(a}\xi_{b)} = 0$. Show that if $R_{ab} = 0$ then the tensor $F_{ab} = \nabla_a \xi_b$ satisfies *Maxwell's vacuum equations* in curved space: $\nabla^a F_{ab} = 0$, $\quad \nabla_{[a}F_{bc]} = 0$.

[7.5] Suppose that F_{ab} satisfies the curved-space Maxwell equations $g^{ab}\nabla_a F_{bc} = 0$, $\quad \nabla_{[a}F_{bc]} = 0$ where g_{ab} is the metric and $\nabla_a g_{bc} = 0$. Show that F_{ab} satisfies the following wave equation: $\nabla^a \nabla_a F_{bc} = \beta R_{bcde}F^{de} + \gamma R^d_{[b}F_{c]d}$ where R_{ab} is the Ricci tensor, defined by $R_{ab} = R_{acb}{}^c$, and β and γ are constants which should be determined.

[7.6] Let g_{ab} be a smooth non-degenerate symmetric space-time metric. Show that there exists a unique torsion-free connection such that $\nabla_a g_{bc} = 0$, given by

$$\Gamma^c_{ab} = \frac{1}{2}g^{cd}(\partial_a g_{bd} + \partial_b g_{ad} - \partial_d g_{ab}).$$

where $\partial_a = \partial/\partial x^a$ and $\nabla_a V_b = \partial_a V_b - \Gamma^c_{ab}V_c$ for all smooth V_a. A new metric $\hat{g}_{ab}$ is defined by $\hat{g}_{ab} = \phi g_{ab}$, where $\phi(x)$ is a smooth scalar field. Compute $\hat{\Gamma}^c_{ab}$ and show that $\Gamma^c_{ab} = \hat{\Gamma}^c_{ab}$ if and only if ϕ is constant. Suppose a vector field ξ_a satisfies

$$\nabla_{(a}\xi_{b)} = \frac{1}{4}g_{ab}g^{cd}\nabla_c\xi_d.$$

Prove that

$$\hat{\nabla}_{(a}\hat{\xi}_{b)} = \frac{1}{4}\hat{g}_{ab}\hat{g}^{cd}\hat{\nabla}_c\hat{\xi}_d$$

where $\hat{\xi}_a = \phi\xi_a$ and $\hat{\nabla}_c$ is the covariant derivative operation associated with $\hat{g}_{ab}$. (The tensor $\hat{g}^{ab}$ is defined by $\hat{g}_{ab}\hat{g}^{bc} = \delta^c_a$.)

[7.7] Show that if $R_{ab} = 0$ then R_{abcd} satisfies the following non-linear wave equation:

$$\nabla^e\nabla_e R_{abcd} + R_{ab}{}^{ef}R_{cdef} + 2R_b{}^e{}_c{}^f R_{aedf} - 2R_a{}^e{}_c{}^f R_{bedf} = 0.$$

[7.8] Let ∇_a be a symmetric (i.e. torsion-free) connection compatible with a given metric g_{ab}. Show that there exists a totally skew tensor ε_{abcd} unique to a constant of proportionality such that $\nabla_a \varepsilon_{bcde} = 0$. Show that there is a unique tensor ε_{abcd} defined by

$$\delta^p_{[a}\delta^q_b\delta^r_c\delta^s_{d]} = \varepsilon_{abcd}\varepsilon^{pqrs}$$

and that ε_{abcd} thus defined satisfies $\nabla_a \varepsilon_{bcde} = 0$.

[7.9] We say that on a space-time M the Ricci tensor R_{ab} is *non-degenerate* if it has everywhere non-vanishing determinant; or, equivalently, if it possesses an inverse, i.e. there exists a tensor S^{bc} such that $R_{ab}S^{bc} = \delta_a^c$. Suppose M has a non-degenerate Ricci tensor. Show that there cannot exist on M a non-trivial vector field K_b satisfying $\nabla_a K_b = 0$.

[7.10] Let us write $g = det(g_{ab})$. Show that for any vector field V^a we have $\nabla_a V^a = (-g)^{\frac{1}{2}} \partial_a (V^a (-g)^{\frac{1}{2}})$ if g_{ab} has Lorentzian signature.

[7.11] Let Y_{ab} be a skew-symmetric tensor that satisfies $\nabla_{(a}Y_{b)c} = 0$. Show that if $R_{ab} = 0$, then F_{ab} defined by $F_{ab} = R_{abcd}Y^{cd}$ satisfies *Maxwell's vacuum equations* in curved space: $\nabla_{[a}F_{bc]} = 0$, $\nabla^a F_{ab} = 0$.

[7.12] Suppose that ξ_a satisfies Killing's equation $\nabla_{(a}\xi_{b)} = 0$. Show as a consequence that $\nabla_a \nabla_b \xi_c = R_{abc}{}^d \xi_d$. The *twist* of ξ^a is a vector field ω_a defined by $\omega_a = \epsilon_{abcd}\xi^b \nabla^c \xi^d$. Show that ω_a is the *gradient* of a scalar if and only if ξ_a is an eigenvector of the Ricci tensor, i.e. $R_{ab}\xi^b = \lambda \xi_a$ for some scalar λ. (Hint: a vector field Ω_b has vanishing *curl* $\nabla_{[a}\Omega_{b]} = 0$ if and only if its dual $\Omega^{abc} = \epsilon^{abcd}\Omega_d$ has vanishing divergence: $\nabla_a \Omega^{abc} = 0$.)

[7.13] Let $\tilde{\nabla}_a$ be a torsion-free connection, and define a new connection ∇_a such that

$$\nabla_a V^b = \tilde{\nabla}_a V^b - \frac{1}{2}T_{ac}^b V^c$$

for any vector field V^b, where $T_{ac}^b = -T_{ca}^b$. Let $\Omega_{ab} = -\Omega_{ba}$ be a non-degenerate two-form with inverse $\hat{\Omega}^{bc}$ such that $\Omega_{ab}\hat{\Omega}^{bc} = \delta_a^c$. Show that there exists a unique choice for T_{ac}^b such that $\nabla_a \Omega_{bc} = 0$.

8   The Lie derivative

HERE we shall introduce a special type of derivation operation, applicable to tensor fields of any rank, which has the property of being *connection independent*.

Let P^a and Q^a be vector fields; then the *Lie derivative* of Q^a with respect to P^a is defined as follows:

$$\pounds_P Q^a = P^b \nabla_b Q^a - Q^b \nabla_b P^a \tag{8.1}$$

where ∇_a is torsion-free. Independence from the specific choice of symmetric connection arises as follows:

$$\begin{aligned} P^b \nabla_b Q^a - Q^b \nabla_b P^a &= P^b \partial_b Q^a + P^b \Gamma^a_{bc} Q^c - Q^b \partial_b P^a - Q^b \Gamma^a_{bc} P^c \\ &= P^b \partial_b Q^a - Q^b \partial_b P^a \end{aligned} \tag{8.2}$$

by use the torsion-free condition $\Gamma^c_{ab} = \Gamma^c_{ba}$. The action on *covariant* vector fields is given by:

$$\pounds_P S_a = P^b \nabla_b S_a + S_b \nabla_a P^b \tag{8.3}$$

which (exercise) can be verified to be independent again of the choice of a symmetric connection. The action on scalar fields is $\pounds_P \phi = P^a \nabla_a \phi$.

We can check that $\pounds_P$ satisfies the *Leibniz property*; thus, e.g., we have:

$$\begin{aligned} \pounds_P(Q^a S_a) &= (\pounds_P Q^a) S_a + Q^a \pounds_P S_a \\ &= (P^b \nabla_b Q^a) S_a - (Q^b \nabla_b P^a) S_a \\ &\quad + Q^a(P^b \nabla_b S_a) + Q^a(S_b \nabla_a P^b) \\ &= P^b \nabla_b(Q^a S_a), \end{aligned} \tag{8.4}$$

as desired.

The action of $\pounds_P$ on tensors of *higher* valence is given according to the following scheme:

$$\pounds_P Q^{ab} = P^c \nabla_c Q^{ab} - Q^{cb} \nabla_c P^a - Q^{ac} \nabla_c P^b \tag{8.5}$$

$$\pounds_P S_{ab} = P^c \nabla_c S_{ab} + S_{cb} \nabla_a P^c + S_{ac} \nabla_b P^c \tag{8.6}$$

and so forth.

As an elementary application of the use of Lie derivatives we note below an *alternative proof of the fundamental theorem of Riemannian geometry.*

Let g_{ab} be a metric tensor ; we wish to establish (again) that there exists a unique symmetric connection ∇_a such that $\nabla_a g_{bc} = 0$.

Proof: Let V^a be any vector field. Then both the expressions $\mathcal{L}_V g_{bc}$ and $\nabla_{[c}(V^a g_{b]a})$ are connection independent, the first being a Lie derivative, the second being an exterior derivative (cf. exercise 4.2). Therefore

$$\mathcal{L}_V g_{bc} + 2\nabla_{[c} V^a g_{b]a} \tag{8.7}$$

is connection independent. Expanding this we deduce that the expression

$$2g_{ab}\nabla_c V^a + V^a(\nabla_a g_{bc} + \nabla_c g_{ab} - \nabla_b g_{ac}) \tag{8.8}$$

is connection independent. Multiplying by $\frac{1}{2}g^{bd}$ we see that

$$\nabla_c V^d + V^a \frac{1}{2}g^{bd}(\nabla_a g_{bc} + \nabla_c g_{ab} - \nabla_b g_{ac}) \tag{8.9}$$

is connection independent, whence

$$\begin{aligned}
&\nabla_c V^d + V^a \frac{1}{2}g^{bd}(\nabla_a g_{bc} + \nabla_c g_{ab} - \nabla_b g_{ac}) \\
&= \partial_c V^d + V^a \{\frac{1}{2}g^{bd}(\partial_a g_{bc} + \partial_c g_{ab} - \partial_b g_{ac})\}.
\end{aligned} \tag{8.10}$$

So if we set $\nabla_a g_{bc} = 0$ (by assumption) it then follows that

$$\nabla_c V^d = \partial_c V^d + V^a \Gamma_{ac}^d, \tag{8.11}$$

where Γ_{ac}^d is the Levi-Civita connection.

The Lie derivative has many uses in relativistic physics—it turns out often to be particularly helpful in reducing rather complicated looking tensorial expressions to simpler formulae, and in doing so giving them a heightened geometrical content.

Exercises for chapter 8

[8.1] Symmetries of a space-time are described by 'Killing vectors', which are directions in which one can move without the 'geometry changing': $\mathcal{L}_K g_{ab} = 0$. Show this is equivalent to K^a satisfying Killing's equation $\nabla_{(a}K_{b)} = 0$.

[8.2] Show that $\mathcal{L}_{[U,V]} = \mathcal{L}_U \mathcal{L}_V - \mathcal{L}_V \mathcal{L}_U$.

[8.3] A space-time has a conserved, isentropic (constant entropy), ideal fluid stress-energy tensor given by $T_{ab} = (\rho + p)u_a u_b - g_{ab}p$ where ρ and p are scalar functions and $u^a u_a = 1$. Suppose the energy density $\rho(x)$ can be expressed as a function of the pressure $p(x)$. Show that the specific enthalpy $f(x)$ of the fluid (cf. equation 3.6.13) is given (up to a constant multiple) by:

$$f(x) = exp\left[\int \frac{dp}{p + \rho}\right].$$

The current vector C_a and vorticity tensor Ω_{ab} of the fluid are defined by

$$C_a = f u_a, \quad \Omega_{ab} = \nabla_{[a} C_{b]},$$

where f is the specific enthalpy. Show that $C^a \Omega_{ab} = 0$ (*Euler's equation*), and that $\mathcal{L}_C \Omega_{ab} = 0$ (*relativistic Helmholtz equation*). The vorticity vector ω^a is defined by $\omega^a = \varepsilon^{abcd} C_b \Omega_{cd}$. Show that $\mathcal{L}_C \omega^a = 0$. Hint: see exercises [3.10] and [3.11].

[8.4] Let Γ^a_{bc} denote a symmetric connection. Show that, although Γ^a_{bc} is not a tensor, nevertheless $\mathcal{L}_\zeta \Gamma^a_{bc}$ transforms as a tensor of valence $\begin{bmatrix} 1 \\ 2 \end{bmatrix}$, and that

$$\mathcal{L}_\zeta \Gamma^a_{bc} = \nabla_b \nabla_c \zeta^a + R_{cbm}{}^a \zeta^m$$

where ∇_a is the covariant derivative associated with Γ^a_{bc}. Show therefore that if $\mathcal{L}_\zeta \Gamma^a_{bc} = 0$ then $K_{ab} = \nabla_{(a} \zeta_{b)}$ is a Killing tensor: $\nabla_{(a} K_{bc)} = 0$. Show furthermore that if $\mathcal{L}_\zeta \Gamma^a_{bc} = 2\delta^a_{(b} \nabla_{c)} \phi$ for some scalar ϕ then $\tilde{K}_{ab} = \nabla_{(a} \zeta_{b)} - 2\phi g_{ab}$ is a Killing tensor.

[8.5] Suppose that M is an n-dimensional Riemannian manifold and that V^a is a vector field such that

$$\mathcal{L}_V g_{ij} = 2\phi g_{ij}$$

for some scalar ϕ. Show that

$$\mathcal{L}_V \Gamma^c_{ab} = \phi_a \delta^c_b + \phi_b \delta^c_a - g_{ab} \phi^c$$

where Γ^a_{bc} is the connection on M and $\phi_a = \nabla_a \phi$. Show that

$$\mathcal{L}_V R_{kji}{}^h = -\delta^h_k \nabla_j \phi_i + \delta^h_j \nabla_k \phi_i - \nabla_k \phi^h g_{ij} + \nabla_j \phi^h g_{ki}.$$

Now suppose M is an *Einstein manifold*, i.e. a manifold with the property that

$$R_{ij} = \frac{1}{n} R g_{ij}$$

where $R_{ij} = R_{iaj}{}^a$ is the Ricci tensor. Show that

$$\nabla_i \phi_j = K \phi g_{ij}.$$

where K is constant. Show that

$$K = \frac{R}{n(n-1)}.$$

[8.6] Verify that

$$X^a \nabla_a Y^b = \frac{1}{2} g^{bc} [(\mathcal{L}_X g_{cd}) Y^d + (\mathcal{L}_Y g_{cd}) X^d \\ - \nabla_c (g_{ab} X^a Y^b)] + \frac{1}{2} \mathcal{L}_X Y^b.$$

This is an explicit formula for the Levi-Civita connection in terms of g_{ab} and metric-independent derivation operations (the Lie derivative and the gradient of a scalar).

[8.7] The Lie derivative can be generalized as follows. Let $S^{ab\cdots c}$ and $T^{ab\cdots d}$ be symmetric tensors of valence m and n respectively. Their *skew product* $P^{b\cdots f}$ is a symmetric tensor of valence $m + n - 1$ defined by

$$P^{b\cdots f} = mS^{r(b\cdots c}\nabla_r T^{d\cdots f)} - nT^{r(b\cdots d}\nabla_r S^{e\cdots f)}.$$

Show that $P^{b\cdots f}$ is independent of the choice of symmetric connection ∇_a. Suppressing indices let us write $P = [S, T]$. Show that $[A, B]$ satisfies the *Jacobi identity*:

$$[[A, B], C] + [[C, A], B] + [[B, C], A] = 0.$$

Let $P \cap Q$ denote the symmetrized outer product of two tensors P and Q. Show that

$$[R, P \cap Q] = [R, P] \cap Q + [R, Q] \cap P.$$

9   Geodesics

9.1   Curves and tangent vectors

LET U_x and $U_{x'}$ be a pair of overlapping coordinate patches in a differentiable manifold M. A curve γ in M is parametrized with parameter u according to the scheme $x^a = x^a(u)$ in the U_x patch, as illustrated in figure 9.1. The *tangent vector* $\xi^a(u)$ to the curve γ is defined by

$$\frac{dx^a(u)}{du} = \xi^a(x(u))$$

in the U_x patch. Suppose we let

$$\frac{dx'^b(u)}{du} = \xi'^b(x'(u))$$

be the tangent vector to γ in the $U_{x'}$ patch. Then, as we shall demonstrate, $\xi^a(x)$ transforms as a *contravariant vector*. The proof runs as follows. We have $x^a(u) = x^a(x'^b(u))$, say, in $U_x \cap U_{x'}$. Thus

$$\frac{dx^a(u)}{du} = \frac{dx'^b(u)}{du} \frac{\partial x^a(x'^b)}{\partial x'^b}.$$

by the chain rule; whence

$$\xi^a(u) = \xi'^b \frac{\partial x^a}{\partial x'^b}$$

and therefore

$$\xi'^b = \frac{\partial x'^b}{\partial x^a} \xi^a,$$

the correct transformation law for a contravariant vector, as we move from patch to patch.

9.2   The absolute derivative along a curve

Let M be a manifold with symmetric connection Γ^c_{ab}, and let $\Gamma^c_{ab}(x^d(u))$ be the restriction of that connection to the curve γ in the patch U_x. Suppose moreover that $V^a(x(u))$ is a contravariant vector field given along the curve γ so that

$$V'^b(u) = \frac{\partial x'^b}{\partial x^a} V^a(u)$$

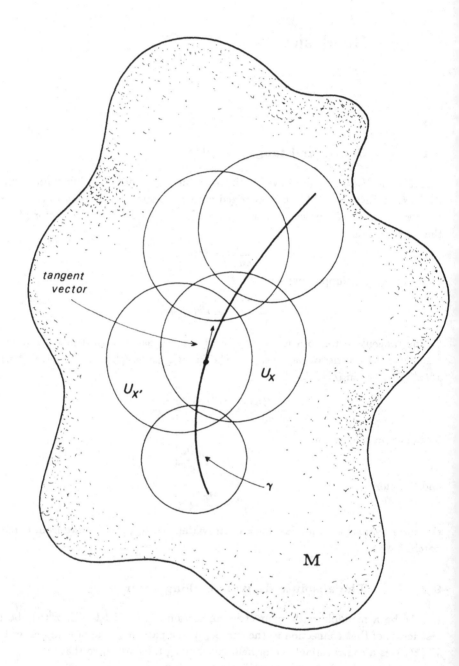

Figure 9.1.   The tangent vector to a curve γ in a manifold M transforms as a contravariant vector ξ^a in transition regions.

in $U_x \cap U_{x'}$. Then $dV^a(u)/du$ along γ does *not* transform properly as a tensor between patches; however the so-called *absolute derivative*, defined by

$$\frac{DV^a}{Du} = \frac{dV^a}{du} + \Gamma^a_{bc}\xi^b V^c$$

does transform properly. The 'bad' terms cancel out, so DV^a/Du also transforms as a covariant vector.

9.3 The geodesic equation

A geodesic on a manifold is defined to be a differentiable curve γ such that the tangent vector $\xi^a(u)$ satisfies

$$\frac{D\xi^a}{Du} = \lambda(u)\xi^a, \quad \xi^a(u) = \frac{dx^a(u)}{du} \tag{9.3.1}$$

along the curve for some function $\lambda(u)$. Thus for a geodesic the absolute derivative of the tangent vector always points in the direction of the tangent vector. More explicitly (9.3.1) can be written as follows:

$$\frac{d^2 x^a(u)}{du^2} + \Gamma^a_{bc}\frac{dx^b}{du}\frac{dx^c}{du} = \lambda(u)\frac{dx^a}{du}. \tag{9.3.2}$$

It is a hypothesis of relativity theory that small 'freely-falling' bodies move along geodesic trajectories. The curvature of space-time (i.e. the gravitational field) makes itself felt through the connection term appearing in the geodesic equation.

9.4 Affine parametrization

Let us consider the effect of a change in parameter $u \to u(\sigma)$ along the curve. We set $u = u(\sigma)$ and $\sigma = \sigma(u)$, so that

$$\frac{dx^a}{du} = \frac{d\sigma}{du}\frac{dx^a}{d\sigma}$$

and

$$\frac{d^2 x^a}{du^2} = \frac{d^2\sigma}{du^2}\frac{dx^a}{d\sigma} + (\frac{d\sigma}{du})^2\frac{d^2 x^a}{d\sigma^2}.$$

Plugging these into the geodesic equation (9.3.2) we get:

$$\frac{d^2\sigma}{du^2}(\frac{dx^a}{d\sigma}) + (\frac{d\sigma}{du})^2\frac{d^2 x^a}{d\sigma^2} + \Gamma^a_{bc}\frac{dx^b}{d\sigma}\frac{dx^c}{d\sigma}(\frac{d\sigma}{du})^2 = \lambda(u)\frac{dx^a}{d\sigma}\frac{d\sigma}{du}.$$

Now suppose $\sigma(u)$ is chosen such that

$$\frac{d^2\sigma}{du^2} = \lambda(u)\frac{d\sigma}{du}.$$

Then we obtain:

$$\frac{d}{du}(\frac{d\sigma}{du}) = \lambda(u)(\frac{d\sigma}{du}),$$

whence

$$\frac{d\sigma}{du} = \int^u exp[\lambda(\alpha)]d\alpha$$

and therefore

$$\sigma(u) = \int^u \int^\beta exp[\lambda(\alpha)]d\alpha d\beta.$$

Thus by choosing $\sigma(u)$ in this way we get:

$$\frac{d^2 x^a}{d\sigma^2} + \Gamma^a_{bc}\frac{dx^b}{d\sigma}\frac{dx^c}{d\sigma} = 0$$

which can be expressed as

$$\frac{D\xi^a}{D\sigma} = 0$$

with

$$\xi^a = \frac{dx^a(\sigma)}{d\sigma}.$$

When these conditions are satisfied σ is called an *affine parameter* along the geodesic. Clearly if σ is an affine parameter then so is $\sigma' = a\sigma + b$ with a, b constants. (This is the only freedom available in the choice of affine parameter.) If $\xi^a\xi_a \neq 0$, then by a suitable affine parametrization we can arrange always that either $\xi^a\xi_a = 1$ or $\xi^a\xi_a = -1$, depending on whether the geodesic is *time-like* or *space-like*. A geodesic for which $\xi^a\xi_a = 0$ is called a *null* geodesic.

Note that for an affinely parametrized geodesic the value of $\xi^a\xi_a$ at any one point along the geodesic is sufficient to determine the value of $\xi^a\xi_a$ at all other points along the geodesic. Thus $\xi^a\xi_a$ is said to be a 'constant of the motion' for the geodesic γ. If T^a is a Killing vector, i.e. a vector field such that $\nabla_{(a}T_{b)} = 0$, then it is straightforward to verify that $T^a\xi_a$ is also constant along γ. If ξ^a is time-like and future-pointing then $T^a\xi_a$ is the 'energy' associated with the vector field ξ_a, and $\xi^a\xi_a$ is the squared mass.

Exercises for chapter 9

[9.1] Let $x^a(s)$ describe a curve, and put $\xi^a = dx^a(s)/ds$. Show that the geodesic equation for ξ^a can be put in *Lagrangian form*

$$\frac{d}{ds}\frac{\partial L}{\partial \dot{x}^a} = \frac{\partial L}{\partial x^a}$$

with $L = \frac{1}{2}g_{ab}\dot{x}^a\dot{x}^b$ and $\dot{x}^a = \xi^a$. Derive this from a suitable variational principle.

[9.2] Suppose a tensor field T^{ab} is of the form $T^{ab} = \rho u^a u^b$ where $\rho(x)$ is a scalar field and $u^a(x)$ satisfies $u^a u_a = 1$. If $\nabla_a T^{ab} = 0$ show that $\nabla_a(\rho u^a) = 0$ and that $(u^a \nabla_a u^b) u^c = (u^a \nabla_a u^c) u^b$.

[9.3] Suppose k^a is null and satisfies $(k^a \nabla_a k^{[b]}) k^{c]} = 0$. Set $\tilde{g}_{ab} = \Phi^2 g_{ab}$ and let $\tilde{\nabla}_a$ be the Levi-Civita connection associated with $\tilde{g}_{ab}$. Show that $(k^a \tilde{\nabla}_a k^{[b]}) k^{c]} = 0$. This shows that the null geodesic condition is *conformally invariant*.

[9.4] Suppose that on a space-time with metric g_{ab} a tensor T_{ab} with $\nabla^a T_{ab} = 0$ is defined by $T_{ab} = (\rho + p) u_a u_b - g_{ab} p$, where ρ and p are scalars and $u^a u_a = 1$. Suppose also that ρ can be expressed as a function of p, and define

$$f(x) = exp \int \frac{dp}{p + \rho(p)}.$$

Set $\hat{g}_{ab} = f^2 g_{ab}$, $\hat{g}^{ab} = f^{-2} g^{ab}$, $\hat{C}_a = f u_a$, $\hat{C}^a = f^{-1} u^a$. Show that with respect to $\hat{g}_{ab}$ and the associated connection $\hat{\nabla}_a$ the current $\hat{C}_a$ is *geodesic*:

$$\hat{C}^a \hat{\nabla}_a \hat{C}^b = 0.$$

[9.5] As a simple model of the universe we assume space-time to be endowed with three Killing vector fields α_a, β_a and γ_a. Show that the vector field V^d defined by

$$V^d = \alpha_a \beta_b \gamma_c \varepsilon^{abcd}$$

satisfies the geodesic equation $V^a \nabla_a V^b = \lambda V^b$ for some scalar λ.

[9.6] Suppose K_{ab} is a symmetric tensor satisfying

$$\nabla_{(a} K_{bc)} = \zeta_{(a} g_{bc)}$$

for some vector ζ_a, which in turn satisfies

$$\nabla_{(a} \zeta_{b)} = \frac{1}{2} k g_{ab},$$

for some constant k. Show that the quantity

$$Q = k s^2 L^2 - 2 s L \zeta_a u^a + K_{ab} u^a u^b$$

is constant along geodesic γ, where u^a is tangent to γ, s is the geodesic parameter and $L = \frac{1}{2} g_{ab} u^a u^b$. Such a K_{ab} is a *homothetic Killing tensor*.

[9.7] A simple, ideal relativistic fluid has unit velocity u^a, specific enthalpy f, and specific entropy S. The *vorticity tensor* Ω_{ab} is defined by $\Omega_{ab} = \nabla_{[a} C_{b]}$ where $C_a = f u_a$ is the current. Let $\omega_a = \epsilon_{abcd} \Omega^{bc} C^d$ denote the *vorticity four-vector*. Show that

$$J^a = f^{-1} u^a \omega^b \nabla_b S$$

has *vanishing divergence*: $\nabla_a J^a = 0$. Hence show that

$$\phi = (\rho + p)^{-1} \omega^a \nabla_a S$$

is a *constant of the fluid's motion*: $u^a \nabla_a \phi = 0$.

10   Geodesic deviation

10.1   Derivation of the geodesic deviation equation

SUPPOSE we have a curve γ in M given by $x^a = x^a(\sigma)$ where σ is a parameter along γ. We write $\xi^a(\sigma) = dx^a(\sigma)/d\sigma$ for the vector tangent to γ. If γ is geodesic, with affine parametrization, then

$$\frac{D\xi^a}{D\sigma} = 0 \Leftrightarrow \frac{d\xi^a}{d\sigma} + \Gamma^a_{bc}\xi^b\xi^c = 0$$

$$\Leftrightarrow \frac{d^2x^a}{d\sigma^2} + \Gamma^a_{bc}\frac{dx^b}{d\sigma}\frac{dx^c}{d\sigma} = 0. \tag{10.1.1}$$

Note that the connection Γ^a_{bc} is specified 'along the path', i.e. $\Gamma^a_{bc}(\sigma) = \Gamma^a_{bc}(x(\sigma))$, where $\Gamma^a_{bc}(x)$ is the connection on the manifold.

Now consider a one-parameter *family* of geodesics $x^a(\sigma, \rho)$, where ρ labels the various geodesics and σ is an affine parameter along the corresponding geodesic for any given value of ρ, as illustrated in figure 10.1. Then the entire family of geodesics is characterized by the equation

$$\frac{\partial^2 x^a(\sigma, \rho)}{\partial\sigma^2} + \Gamma^a_{bc}(x(\sigma, \rho))\frac{\partial x^b}{\partial\sigma}\frac{\partial x^c}{\partial\sigma} = 0 \tag{10.1.2}$$

which must hold for all ρ, σ in some specified range. For each fixed value of ρ we get the geodesic equation corresponding to that value of ρ. Let γ_1 and γ_2 be two neighbouring curves, specified by slightly different values of ρ, so:

$$\gamma_1 \sim x^a(\sigma, \rho), \quad \gamma_2 \sim x^a(\sigma, \rho + \varepsilon)$$

with ε small. Define $\eta^a(\sigma, \rho, \varepsilon)$ by

$$\varepsilon\eta^a(\sigma, \rho, \varepsilon) = x^a(\sigma, \rho + \varepsilon) - x^a(\sigma, \rho).$$

Then in the limit $\varepsilon \to 0$ we have

$$\eta^a(\sigma, \rho, 0) = \eta^a(\sigma, \rho) = \frac{\partial x^a(\sigma, \rho)}{\partial\rho}.$$

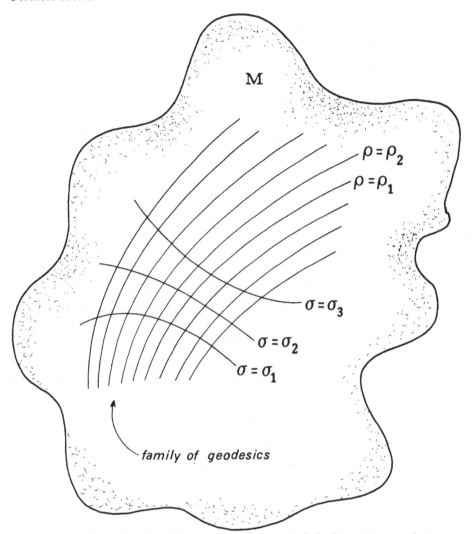

Figure 10.1   A family of neighbouring geodesics: ρ labels the various geodesics, σ and is an affine parameter along the corresponding geodesic for a given value of ρ.

Let γ be the geodesic corresponding to some fixed value of ρ, say $\rho = \rho_0$. Put $\eta^a(\sigma) = \eta^a(\sigma, \rho_0)$. Then $\eta^a(\sigma)$ is a contravariant vector field along γ which measures the infinitesimal displacement of neighbouring geodesics from γ as σ varies. The way in which $\eta^a(\sigma)$ varies with respect to σ is determined by the following fundamental relation, called the *equation of geodesic deviation*:

$$\frac{D^2 \eta^a(\sigma)}{D\sigma^2} = R_{bcd}{}^a(x(\sigma))\xi^b \eta^c \xi^d. \qquad (10.1.3)$$

To establish this relation we differentiate (10.1.2) with respect to ρ to get:

$$-\frac{\partial^2}{\partial\sigma^2}\frac{\partial}{\partial\rho}x^a(\rho,\sigma) = \frac{\partial}{\partial\rho}[\Gamma^a_{bc}(x(\sigma,\rho))\frac{\partial x^b}{\partial\sigma}\frac{\partial x^c}{\partial\sigma}]$$

$$= \frac{\partial x^b}{\partial\sigma}\frac{\partial x^c}{\partial\sigma}\frac{\partial}{\partial\rho}\Gamma^a_{bc}(x(\sigma,\rho)) + 2\Gamma^a_{bc}(x(\sigma,\rho))\frac{\partial x^b}{\partial\sigma}\frac{\partial^2 x^c}{\partial\sigma\partial\rho}.$$

Now set $\rho = \rho_0$, and put

$$\left[\frac{\partial x^a(\sigma,\rho)}{\partial\rho}\right]_{\rho=\rho_0} = \eta^a(\sigma),$$

and

$$\left[\frac{\partial x^a(\sigma,\rho)}{\partial\sigma}\right]_{\rho=\rho_0} = \xi^a(\sigma).$$

Noting that

$$\frac{\partial}{\partial\rho}\Gamma^a_{bc}(x(\sigma,\rho)) = (\frac{\partial x^d}{\partial\rho})\partial_d\Gamma^a_{bc}(x)$$

we obtain

$$-\frac{d^2\eta^a(\sigma)}{d\sigma^2} = \partial_d\Gamma^a_{bc}\eta^d\xi^b\xi^c + 2\Gamma^a_{bc}\xi^b\frac{d\eta^c}{d\sigma}.$$

Making a few cosmetic alterations in this relation (by changing indices) we get:

$$\frac{d^2\eta^a}{d\sigma^2} = -\partial_c\Gamma^a_{bd}\xi^b\eta^c\eta^d - 2\Gamma^a_{bc}\xi^b\frac{d\eta^c}{d\sigma}. \qquad (10.1.4)$$

This formula is the first step in the calculation, and will be used shortly. Now recall the definition of the *absolute derivative*. For any vector field $V^a(x)$ its absolute derivative along the curve γ is

$$\frac{DV^a}{D\sigma} = \frac{dV^a(\sigma)}{d\sigma} + \Gamma^a_{bc}V^b\xi^c.$$

Thus the absolute derivative of $\eta^a(\sigma)$ along γ is

$$\frac{D\eta^a}{D\sigma} = \frac{d\eta^a}{d\sigma} + \Gamma^a_{pq}\xi^p\eta^q.$$

The second absolute derivative is then:

$$\frac{D^2\eta^a}{D\sigma^2} = \frac{d}{d\sigma}(\frac{D\eta^a}{D\sigma}) + \Gamma^a_{rs}\frac{D\eta^r}{D\sigma}\xi^s$$

$$= \frac{d}{d\sigma}(\frac{d\eta^a}{d\sigma} + \Gamma^a_{pq}\xi^p\eta^q) + \Gamma^a_{rs}(\frac{d\eta^r}{d\sigma} + \Gamma^r_{pq}\xi^p\eta^q)\xi^s$$

$$= \frac{d^2\eta^a}{d\sigma^2} + \partial_d\Gamma^a_{pq}\xi^d\xi^p\eta^q + \Gamma^a_{pq}\frac{d\xi^p}{d\sigma}\eta^q + \Gamma^a_{pq}\xi^p\frac{d\eta^q}{d\sigma}$$

$$+ \Gamma^a_{rs}\frac{d\eta^r}{d\sigma}\xi^s + \Gamma^a_{rs}\Gamma^r_{pq}\xi^p\eta^q\xi^s.$$

In term 3 in the expression above we can use the geodesic equation

$$\frac{D\xi^a}{D\sigma} = \frac{d\xi^a}{d\sigma} + \Gamma^a_{rs}\xi^r\xi^s = 0$$

in order to put

$$\frac{d\xi^a}{d\sigma} = -\Gamma^a_{rs}\xi^r\xi^s.$$

Terms 4 and 5 are identical. So we have:

$$\frac{D^2\eta^a}{D\sigma^2} = \frac{d^2\eta^a}{d\sigma^2} + \partial_d\Gamma^a_{pq}\xi^d\xi^p\eta^q$$
$$+ 2\Gamma^a_{pq}\xi^p\frac{d\eta^q}{d\sigma} + \Gamma^a_{rs}\Gamma^r_{pq}\xi^p\eta^q\xi^s - \Gamma^a_{pq}\Gamma^p_{rs}\xi^r\xi^s\eta^q.$$

So finally with some index re-arrangement we get

$$\frac{D^2\eta^a}{D\sigma^2} = \frac{d^2\eta^a}{d\sigma^2} + \partial_b\Gamma^a_{cd}\xi^b\eta^c\xi^d$$
$$+ 2\Gamma^a_{bc}\xi^b\frac{d\eta^c}{d\sigma} + (\Gamma^a_{br}\Gamma^r_{cd} - \Gamma^a_{cr}\Gamma^r_{bd})\xi^b\eta^c\xi^d. \tag{10.1.5}$$

The last step of the calculation is to insert expression (10.1.4) for $d^2\eta^a/d\sigma^2$ in (10.1.5) above. The $2\Gamma^a_{bc}\xi^b\frac{d\eta^c}{d\sigma}$ terms cancel, leaving:

$$\frac{D^2\eta^a}{D\sigma^2} = (\partial_b\Gamma^a_{cd} - \partial_c\Gamma^a_{bd})\xi^b\eta^c\xi^d$$
$$+ (\Gamma^a_{br}\Gamma^r_{cd} - \Gamma^a_{cr}\Gamma^r_{bd})\xi^b\eta^c\xi^d$$
$$= R_{bcd}{}^a\xi^b\eta^c\xi^d \tag{10.1.6}$$

(cf. the definition of $R_{abc}{}^d$ in chapter 6).

10.2   Commutators of absolute derivatives

Let $x^a(\alpha, \beta)$ be a 2-surface S parametrized by α and β, as illustrated in figure 10.2. For each fixed value of β, say $\beta = \beta_0$, we have a curve $x^a(\alpha, \beta_0)$ parametrized by α. Let $V^a(\alpha, \beta) = V^a(x(\alpha, \beta))$ be a vector field defined on the surface S. Then we can define the absolute derivatives $D/D\alpha$ and $D/D\beta$ as before, i.e. by

$$\frac{DV^a}{D\alpha} = \frac{dV^a}{d\alpha} + \Gamma^a_{bc}V^bA^c, \tag{10.2.1}$$

where $A^c = \partial x^c/\partial\alpha$, and by

$$\frac{DV^a}{D\beta} = \frac{dV^a}{d\beta} + \Gamma^a_{bc}V^bB^c, \tag{10.2.2}$$

where $B^c = \partial x^c/\partial\beta$. We are led to the following result:

(10.2.3) Theorem.

$$(\frac{D^2}{D\alpha D\beta} - \frac{D^2}{D\beta D\alpha})V^d(\alpha, \beta) = R_{abc}{}^d A^a B^b V^c.$$

Proof: One way is simply to write the relevant terms out explicitly, using Γ's and so on, and using identities such as

$$\frac{\partial}{\partial\alpha}\Gamma^a_{bc} = (\partial_d\Gamma^a_{bc})A^d.$$

The formula of the theorem then follows after a straightforward calculation. Alternatively, and rather more geometrically, one can proceed as follows. We require two lemmas.

(10.2.4) Lemma. *If $V^a(x(\alpha))$ is any vector field defined on the curve $x^a = x^a(\alpha)$, then $DV^b/D\alpha = A^a\nabla_a V^b$, where A^a is the tangent to the curve, $A^a = dx^a(\alpha)/d\alpha$.*

Proof:

$$
\begin{aligned}
\frac{DV^b(x(\alpha))}{D\alpha} &= \frac{dV^b}{d\alpha} + \Gamma^b_{ad}A^aV^d \\
&= \frac{dx^a}{d\alpha}\frac{dV^b}{dx^a} + \Gamma^b_{ad}A^aV^d \\
&= A^a\frac{dV^b}{dx^a} + \Gamma^b_{ad}A^aV^d \\
&= A^a\left(\frac{dV^b}{dx^a} + \Gamma^b_{ad}V^d\right) \\
&= A^a\nabla_a V^b.
\end{aligned}
$$

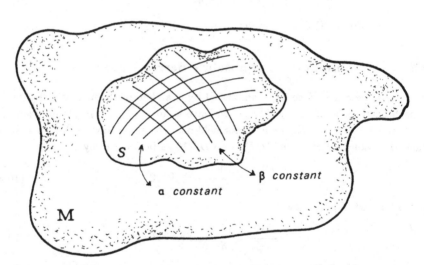

Figure 10.2. A two-surface S parametrized by α and β.

(10.2.5) Lemma. *If $A^a = \partial x^a(\alpha,\beta)/\partial\alpha$ and $B^a = \partial x^a(\alpha,\beta)/\partial\beta$, then $A^a\nabla_a B^b - B^a\nabla_a A^b = 0$.*

Proof: We know that $A^a\nabla_a B^b - B^a\nabla_a A^b$ is connection independent. Thus $A^a\nabla_a B^b - B^a\nabla_a A^b = A^a\partial_a B^b - B^a\partial_a A^b$. But $A^a\partial_a = \partial/\partial\alpha$ and $B^a\partial_a = \partial/\partial\beta$. So

$$
A^a\partial_a B^b - B^a\partial_a A^b = \frac{\partial B^b}{\partial\alpha} - \frac{\partial A^b}{\partial\beta} = \frac{\partial^2 x^b}{\partial\alpha\partial\beta} - \frac{\partial^2 x^b}{\partial\beta\partial\alpha} = 0.
$$

Now, back to the proof of the theorem. By lemma (10.2.3) formula (10.1.3) can be rephrased as follows:

$$A^a \nabla_a (B^b \nabla_b V^d) - B^a \nabla_a (A^b \nabla_b V^d) = R_{abc}{}^d A^a B^b V^c$$

with $A^a = \partial x^a / \partial \alpha$ and $B^a = \partial x^a / \partial \beta$. By the Leibniz rule, however, we have:

$$(A^a \nabla_a B^b \nabla_b - B^a \nabla_a A^b \nabla_b) V^d = A^a B^b (\nabla_a \nabla_b - \nabla_b \nabla_a) V^d + (A^a \nabla_a B^b - B^a \nabla_a A^b) \nabla_b V^d$$

from which the desired result follows immediately, by use of the Ricci identities and lemma (10.2.4).

10.3   Simplified derivation of the geodesic deviation equation

As before, let $x^a(\sigma, \rho)$ be a family of affinely parametrized geodesics, with each geodesic given by some choice of constant ρ. Define

$$\xi^a = \frac{\partial x^a(\sigma, \rho)}{\partial \sigma}, \qquad \eta^a = \frac{\partial x^a(\sigma, \rho)}{\partial \rho},$$

so $D\xi^a / D\sigma = 0$, the geodesic equation with affine parametrization. By lemma (10.2.3) we have

$$\frac{D\eta^a}{D\sigma} = \frac{D\xi^a}{D\rho}$$

and thus

$$\frac{D^2 \eta^a}{D\sigma^2} = \frac{D^2 \xi^a}{D\sigma D\rho}.$$

$$= \frac{D^2 \xi^a}{D\sigma D\rho} - \frac{D^2 \xi^a}{D\rho D\sigma}$$

$$= R_{bcd}{}^a \xi^b \eta^c \xi^d,$$

as desired.

10.4   Parallel propagation

Let $x^a(\sigma)$ be a curve with tangent vector $\xi^a(\sigma) = \partial x^a(\sigma)/\partial \sigma$. Let $V^a(\sigma_0)$ be a vector assigned to a point $x^a = x^a(\sigma_0)$ on the curve. How can we define a vector field $V^a(\sigma)$ along the curve (with $V^a(\sigma)|_{\sigma_0} = V^a(\sigma_0)$) so that each vector $V^a(\sigma)$ can be regarded in an appropriate sense as 'parallel' to the original vector $V^a(\sigma_0)$?

Answer: by requiring $DV^a(\sigma)/D\sigma = 0$ along the curve. Note that

$$\frac{DV^a(\sigma)}{D\sigma} = 0 \Longrightarrow \frac{dV^a(\sigma)}{d\sigma} + \Gamma^a_{bc}(\sigma)\xi^b V^c = 0.$$

Since $\Gamma^a_{bc}(\sigma)$ and $\xi^b(\sigma)$ are definite given functions of σ, the equations of parallel propagation form a system of the form

$$\frac{dV^a}{d\sigma} + M^a_b V^b = 0$$

with $M_b^a(\sigma) = \Gamma_{bc}^a \xi^c$, i.e. four ordinary first order linear differential equations in four unknowns. Assuming reasonable smoothness conditions, there exists a unique solution $V^a(\sigma)$ taking on the correct value at σ_0.

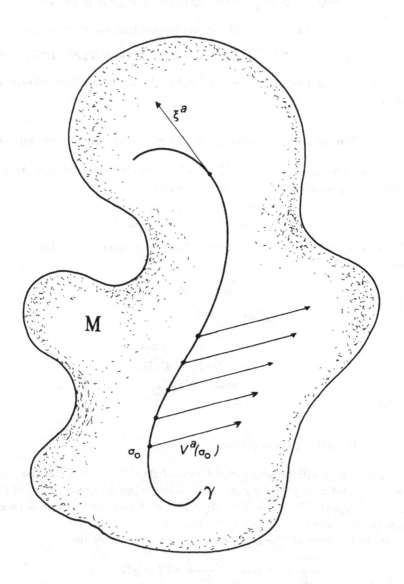

Figure 10.3.   A vector with value $V^a(\sigma_0)$ at $x^a(\sigma_0)$ is parallelly propagated along a curve γ. The curve is given by $x^a(\sigma)$, and has tangent vector $\xi^a(\sigma)$.

Note that a geodesic, affinely parametrized, has the property that its tangent vector is parallelly propagated along the curve. More generally, if two points A and B are connected by a pair of distinct curves α and β, and a vector V^a is parallelly propagated from A to B, then the result will differ according as to whether V^a has been propagated along α or β.

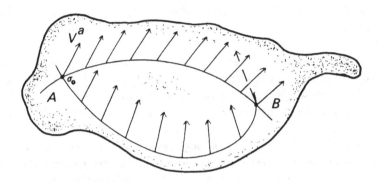

Figure 10.4.   A vector V^a when transported in a Riemannian manifold from A to B by parallel propagation along two distinct paths will generally arrive at B with a result that depends on the choice of path: the result is path independent only if the curvature vanishes.

Exercises for chapter 10

[10.1] A symmetric tensor field K_{ab} is said to be a Killing tensor if $\nabla_{(a}K_{bc)} = 0$ where ∇_a is the torsion-free metric connection. If L_a and M_a are Killing vectors show that $L_{(a}M_{b)}$ is a Killing tensor. Suppose t^a is the tangent vector to an affinely parametrized geodesic curve γ and K_{ab} is a Killing tensor. Show that $K_{ab}t^a t^b$ is constant along γ. Show that if K_{ab} is a Killing tensor then

$$\nabla_{(a}\nabla_b K_{c)d} = -R_{d(ab}{}^e K_{c)e}.$$

By use of this result show that if t^a is tangent to an affinely parametrized geodesic then $\eta_a = K_{ab}t^b$ satisfies the geodesic deviation equation $D^2\eta_a = R_{abcd}t^b\eta^c t^d$, where the operator D is defined by $D = t^a\nabla_a$.

[10.2] Show that in n dimensions the number of independent components of R_{abcd} is $\frac{1}{12}n^2(n^2-1)$, so that in spacetime we have 20, while on a surface there is one.

[10.3] With the usual coordinates a sphere of radius r has metric

$$ds^2 = r^2(d\theta^2 + \sin^2\theta d\phi^2).$$

(i) By use of (6.1.6) find the single independent component of the Riemann tensor and show that the curvature scalar (R_a^a) is $2/r^2$ i.e. simply twice the usual Gaussian curvature. (ii) The geodesics are clearly the great circles. By looking at the relative acceleration of neighbouring great circles and comparing with the equation of geodesic deviation arrive at the same answer as (i).

[10.4] Two very small vector fields U^a and V^a are used to construct a very small parallelogram: $U^a, V^a, -U^a, -V^a$. Parallel transport a vector A^a around this path and show that upon return it has changed by $\delta A^d = R_{abc}{}^d U^a V^b A^c$. Deduce that its length has not changed and hence that it has simply been rotated.

[10.5] A vector field V^a is tangent to a congruence of time-like geodesics, parametrized by proper time s. A vector η^a is propagated along a geodesic γ of the congruence in such a way that $\mathcal{L}_V\eta^a = 0$. Show that η^a satisfies the equation of geodesic deviation $D^2\eta^a = R_{bcd}{}^a V^b \eta^c V^d$ where $D = V^a\nabla_a$. Show that if η^a is initially orthogonal to V^a then it remains orthogonal to V^a. Suppose that a space-time has a curvature tensor R_{abcd} such that

$$R_{abcd} = \frac{1}{12}R(g_{ac}g_{bd} - g_{ad}g_{bc}).$$

Show that for such a space R is necessarily constant. Consider the geodesic equation in such a space. Show that

$$D^2\eta^a = \frac{1}{12}R\eta^a$$

if η^a is orthogonal to V^a. Show that

$$k = D\eta_a D\eta^a - \frac{1}{12}R\eta_a\eta^a$$

is *constant* along γ. Define $L = -\eta_a\eta^a$. Show that

$$D^2L = \frac{1}{3}RL - 2k.$$

Show that if $R > 0$ the solution that tends to zero as $s \to -\infty$ is given by $L = A\exp(s\sqrt{R/3})$ where A is a constant.

11   Differential forms

Various attempts have been made to set up a standard terminology in this branch of mathematics involving only the vectors themselves and not their components, analogous to that of vectors in vector analysis. This is highly expedient in the latter but very cumbersome for the much more complicated framework of the tensor calculus. In trying to avoid continual reference to the components we are obliged to adopt an endless profusion of names and symbols in addition to an intricate set of rules for carrying out calculations, so that the balance of advantage is considerably on the negative side. An emphatic protest must be entered against these orgies of formalism which are threatening the peace of even the technical scientist.

—H. Weyl (Space, Time, Matter)

11.1   A fresh look at anti-symmetric tensors

WE have introduced local differential geometry in a notation that makes great use of indices. This is the classical route and it does have a great deal of merit. There is a parallel development in an *index free* notation that is more generally used by pure mathematicians. The different approaches have their separate advantages and drawbacks: a calculation with indices may be cumbersome and sprawling; conversely an index-free notation may labour what is easily written with indices. On the whole, the index notation is best: but exceptions to this principle do exist.

In this chapter, we discuss the algebra and calculus of differential forms in an index-free notation to give a flavour of this approach, also especially to describe a simple and powerful technique for calculating the Riemann tensor of a given metric.

From chapter 4 we know that a *form* or *differential form* is a totally skew covariant tensor:

$$\omega_{a\cdots b} = \omega_{[a\ldots b]}.$$

The number of indices is the *type* of the form, and a form of type p is called a *p-form*. Note that p must be no greater than n, the dimension of the space; otherwise the form automatically vanishes. By convention, a function is a 0-form. The index-free notation here is particularly natural since the indices serve almost no purpose: one simply writes the kernel letter and remembers the type.

Forms at a point generate an algebra. Clearly the forms of a given type form a real vector space: but there is also a natural product, called the *wedge product*. If α, β are forms of type p, q respectively, then their wedge product

$$\gamma_{a\ldots\ldots d} = \alpha_{[a\ldots\ldots b}\beta_{c\ldots\ldots d]} \tag{11.1.1}$$

is a form of type $p + q$. In the index-free notation, we write (11.1.1) as $\gamma = \alpha \wedge \beta$. We easily find

$$i)\quad (\alpha_1 + \alpha_2) \wedge \beta = \alpha_1 \wedge \beta + \alpha_2 \wedge \beta \qquad (11.1.2)$$

$$ii)\quad (\alpha \wedge \beta) \wedge \gamma = \alpha \wedge (\beta \wedge \gamma)$$

$$iii)\quad \alpha \wedge \beta = (-1)^{pq}\beta \wedge \alpha \text{ where } \alpha \text{ and } \beta \text{ are of types } p \text{ and } q.$$

The algebra of forms is known as the *Grassmann algebra*. If we now imagine fields of forms, then there is a natural derivative operation that raises the type by one. If α is a p-form then

$$\nabla_{[a}\alpha_{b....c]} \qquad (11.1.3)$$

is the *exterior derivative* of α written $d\alpha$. Here (11.1.2) may be calculated with any torsion-free connection, and then we know that it is independent of connection. The exterior derivative is part of the *manifold structure* and is given *prior to any choice of connection*. In this sense the algebra and calculus of forms on a differentiable manifold can be regarded as a fundamental or 'primitive' structure, preceeding the use of a connection. From (11.1.3) and the definition of wedge product (11.1.1) we find

$$(i)\quad d(d\alpha) = d^2\alpha = 0 \quad \text{for all } \alpha \qquad (11.1.4)$$

$$(ii)\quad d(\alpha \wedge \beta) = d\alpha \wedge \beta + (-1)^p \alpha \wedge d\beta$$

where α is a p-form. A form α for which $d\alpha$ is zero is said to be *closed*; while if $\alpha = d\beta$ for some β, we say that α is *exact*. The extent to which closed forms are globally exact on a region of a manifold M is a topological question of great interest, which leads to a topological classification of manifolds.

The exterior derivatives dx^a of the coordinate functions x^a on a region of a manifold form a basis for the 1-forms, and all possible wedge products of these give a basis for all forms. Thus any p-form α can be written

$$\alpha = \alpha_{a...b}dx^a \wedge \cdots \wedge dx^b$$

and the exterior derivative can be simply taken as

$$d\alpha = \frac{\partial \alpha_{a...b}}{\partial x^c}dx^c \wedge dx^a \wedge \ldots \wedge dx^b. \qquad (11.1.5)$$

11.2   Cartan's method

If a manifold has a choice of metric then we may choose a basis of 1-forms with the property that all the inner products of all the basis elements are constant. In particular, we might choose an orthonormal basis, but we wish to allow also for slightly more general possibilities.

Suppose $\{\theta^i, i = 0, 1, 2, 3\}$ is the basis, so that the metric is

$$ds^2 = \mu_{ij}\theta^i\theta^j \tag{11.2.1}$$

where μ_{ij} is a matrix of constants. Here i is a label to say which basis element is involved, and not in any sense a space-time index. Each θ^i is a 1-form and so can be written $\theta^i_a dx^a$ where the index a is a space-time index. The two types of index will be kept separate as much as possible!

There is a dual basis of vector fields $\{e_i, i = 0, 1, 2, 3\}$. Given the metric, we know the Levi-Civita connection, and thus we may differentiate each θ^i along each e_j. The result is a 1-form and so can be expressed in terms of the basis:

$$e^c_j\nabla_c\theta^i_a = -\Gamma^i{}_{jk}\theta^k_a. \tag{11.2.2}$$

The scalars $\Gamma^i{}_{kj}$ are known as *Ricci rotation coefficients* and evidently contain the information of the connection. Also, since the θ^i have constant inner products μ^{ij}, so $\theta^j_b e^c_j = \delta^c_b$, if we contract both sides of (11.2.2) with $g^{ab}\theta^m_b$ we find

$$\Gamma^{(i}{}_{kj}\mu^{m)k} = 0. \tag{11.2.3}$$

We define the so-called connection 1-forms $\omega^i{}_k$ by

$$\omega^i{}_k = \Gamma^i{}_{kj}\theta^j. \tag{11.2.4}$$

Evidently, these also contain the information of the connection and by (11.2.3) they are skew-symmetric if an index is raised by μ^{ij}:

$$\omega^{(i}{}_k\mu^{m)k} = 0. \tag{11.2.5}$$

Now from (11.2.2), since the e_i form a basis dual to the θ^j, so $\theta^j_b e^c_j = \delta^c_b$, if we contract both sides with θ^j_b we find:

$$\nabla_b\theta^i_a = -\omega^i_{kb}\theta^k_a.$$

And if we skew this on $[ab]$, then we revert to the index-free notation with

$$d\theta^i = -\omega^i{}_k \wedge \theta^k. \tag{11.2.6}$$

Given the basis θ^i explicitly, we can solve (11.2.5) and (11.2.6) for $\omega^i{}_k$. If we change the original choice of basis, say by a linear transformation

$$\theta^i \to \hat{\theta}^i = L^i_j\theta^j \tag{11.2.7}$$

while preserving the normalization, then from (11.2.6) we find

$$\omega^i{}_j \to \hat{\omega}^i{}_j = (L^i_m\omega^m{}_n - dL^i_n)\tilde{g}^n_j \tag{11.2.8}$$

where $\tilde{L}^n_j$ is the inverse of L^i_n so $L^i_n\tilde{L}^n_j = \delta^i_j$. The inhomogeneous term in this is the counterpart of the inhomogeneous term in the transformation law for the Christoffel symbols, and in the same way it indicates that the quantity involved is not a tensor.

We expect curvature to appear as second derivatives of θ^i, or as derivatives of $\omega^i{}_j$. Furthermore, the curvature is a tensor. Consider the combination

$$\Omega^i{}_j = d\omega^i{}_j + \omega^i{}_k \wedge \omega^k{}_j. \qquad (11.2.9)$$

Under (11.2.7), with the aid of (11.2.8), we find that $\Omega^i{}_j$ transforms as

$$\Omega^i{}_j \to \hat{\Omega}^i{}_j = L^i_m \Omega^m{}_n \tilde{L}^n{}_j. \qquad (11.2.10)$$

Thus $\Omega^i{}_j$, which we can think of as a matrix of 2-forms, does indeed transform homogeneously, as a tensor should. In fact, a straightforward calculation, which we give as an exercise, leads to

$$\Omega^i{}_j = \frac{1}{2} R_{mnj}{}^i \theta^m \wedge \theta^n \qquad (11.2.11)$$

where $R_{mnj}{}^i$ are the components of the Riemann tensor with respect to the chosen basis θ^i.

Since d^2 vanishes (property (i) in 11.1.4) we obtain a pair of identities from (11.2.6) and (11.2.9). From d applied to (11.2.6) we eventually find:

$$\Omega^i{}_j \wedge \theta^j = 0, \qquad (11.2.12)$$

which is equivalent to proposition (6.3.1). And from d applied to (11.2.9) we find:

$$d\Omega^i{}_j - \Omega^i{}_k \wedge \omega^k{}_j + \omega^i{}_k \wedge \Omega^k{}_j = 0. \qquad (11.2.13)$$

This is the *Bianchi identity* (proposition 6.4.1), reappearing in the notation of differential forms.

11.3   The example of spherically symmetric geometries

In chapter 15 we shall require the Ricci tensor of the metric

$$ds^2 = e^{2\lambda} dt^2 - e^{2\mu} dr^2 - r^2 (d\theta^2 + sin^2\theta d\phi^2) \qquad (11.3.1)$$

where λ and μ are a pair of functions of r. This is the general metric form for a time-independent spherically symmetric geometry—it describes, for example, the geometry outside a spherically symmetric star, once λ and μ have been appropriately determined. We shall calculate the Ricci tensor here as an example of the use of the Cartan method. We take the basis of forms

$$\theta^0 = e^\lambda dt, \quad \theta^1 = e^\mu dr, \quad \theta^2 = r d\theta, \quad \theta^3 = r \sin\theta d\phi \qquad (11.3.2)$$

so that in (11.2.1) μ_{ij} is the usual Minkowski metric $diag(+1, -1, -1, -1)$. From (11.2.5) this means that the connection 1-forms have the symmetry

$$\omega^0{}_i = \omega^i{}_0 \quad i = 1, 2, 3$$

$$\omega^i{}_j = -\omega^j{}_i \quad i, j = 1, 2, 3, \; i \neq j,$$

the rest vanishing. First we calculate the exterior derivatives of the basis by use of (11.1.5). We get:

$$d\theta^0 = \lambda' e^\lambda dr \wedge dt = \lambda' e^{-\mu} \theta^1 \wedge \theta^0$$

$$d\theta^1 = 0$$

$$d\theta^2 = dr \wedge d\theta = \frac{1}{r} e^{-\mu} \theta^1 \wedge \theta^2$$

$$d\theta^3 = \sin\theta dr \wedge d\phi + r\cos\theta d\theta \wedge d\phi = \frac{1}{r} e^{-\mu} \theta^1 \wedge \theta^3 + \frac{1}{r}\cot\theta \theta^2 \wedge \theta^3$$

where prime stands for differentiation with respect to r. From (11.2.6) we equate these to $-\omega^i{}_j \wedge \theta^j$ and find

$$\omega^0{}_1 = \lambda' e^{-\mu} \theta^0 = \omega^1{}_0$$

$$\omega^0{}_2 = \omega^0{}_3 = \omega^2{}_0 = \omega^3{}_0 = 0$$

$$\omega^1{}_2 = -\frac{1}{r} e^{-\mu} \theta^2 = -\omega^2{}_1$$

$$\omega^1{}_3 = -\frac{1}{r} e^{-\mu} \theta^3 = -\omega^3{}_1$$

$$\omega^2{}_3 = -\frac{1}{r}\cot\theta \theta^3 = -\omega^3{}_2.$$

This part of the process involves solving a number of simultaneous linear equations; but since the solution is unique, judicious guessing can eliminate much of the work. Next we calculate the curvature 2-forms from (11.2.9):

$$\Omega^0{}_1 = -(\lambda'' - (\lambda')^2 - \lambda'\mu')e^{-2\mu} \theta^0 \wedge \theta^1$$

$$\Omega^0{}_2 = -\frac{1}{r}\lambda' e^{-2\mu} \theta^0 \wedge \theta^2$$

$$\Omega^0{}_3 = -\frac{1}{r}\lambda' e^{-2\mu} \theta^0 \wedge \theta^3$$

$$\Omega^1{}_2 = \frac{1}{r}\mu' e^{-2\mu} \theta^1 \wedge \theta^2$$

$$\Omega^1{}_3 = \frac{1}{r}\mu' e^{-2\mu} \theta^1 \wedge \theta^3$$

$$\Omega^2{}_3 = \frac{1}{r^2}(1 - e^{-2\mu}) \theta^2 \wedge \theta^3.$$

Finally, from this table, and with the aid of (11.2.11) we read off the Riemann tensor components:

$$R_{011}{}^0 = -(\lambda'' - (\lambda')^2 - \lambda'\mu')e^{-2\mu}$$

$$R_{022}{}^0 = -\frac{1}{r}\lambda' e^{-2\mu} = R_{033}{}^0$$

$$R_{122}{}^1 = \frac{1}{r}\mu' e^{-2\mu} = R_{133}{}^1$$

$$R_{233}{}^2 = \frac{1}{r^2}(1 - e^{-2\mu}).$$

Riemann tensor components that cannot be obtained from these by simple symmetry equations are all zero. For chapter 15 we require the Ricci tensor, which we get by a trace on the Riemann tensor:

$$R_{ab} = R_{acb}{}^c$$

so that

$$R_{00} = e^{-2\mu}(\lambda'' - (\lambda')^2 - \lambda'\mu' + 2\frac{\lambda'}{r})$$

$$R_{11} = e^{-2\mu}(\lambda'' - (\lambda')^2 - \lambda'\mu' - 2\frac{\mu'}{r})$$

$$R_{22} = -\frac{1}{r^2}(1 - e^{-2\mu}) + \frac{1}{r}(\lambda' - \mu')e^{-2\mu} = R_{33}.$$

This is a much simpler technique for calculating Riemann and Ricci tensors than working out the Christoffel symbols and using them. For large classes of metrics the appropriate choice of basis is clear and the exterior derivatives are easy to calculate. Finding the connection 1-forms involves a bit of equation solving, and then calculating the curvature 2-forms is again automatic.

The Cartan calculus can, of course, be pursued to much greater depth, and has numerous applications in relativity theory and indeed differential geometry generally—perhaps the lesson to be learned here is that once the theory of a particular construction is set up (e.g. the theory of the Riemann tensor), then rather characteristically someone else may come along later (in this case Cartan) and devise a special and much faster method of carrying out practical calculation. As a consequence the theory as a whole is studded here and there with special 'modules' (theories within the theory) that are brought into play as necessary.

Exercises for chapter 11

[11.1] Check the assertions in (11.1.2) and (11.1.4).

[11.2] Verify (11.2.11), (11.2.12) and (11.2.13) and see that these last two are the Bianchi identities.

[11.3] Show that, in a simply connected region in R^n, a closed 1-form is exact. (The proof follows the proof of the familiar fact in R^3 that if curl $\mathbf{V} = 0$ then $\mathbf{V} = grad\, f$ for some f. Along the way, notice that $\mathbf{V}$ is curl-free if it is a closed 1-form.)

[11.4] Using spherical polar coordinates in R^3 with the origin removed show that the 2-form $\alpha = sin\theta d\theta \wedge d\phi$ is closed, but is not exact. (The point of removing the origin is that otherwise α is singular there. This is an instance of a property of the space, namely a hole in it, being 'detected' by the existence of a closed but non-exact form on the space.)

[11.5] A 1-form V is *hypersurface-orthogonal* if it is proportional to the normal to a family of hypersurfaces. If the hypersurfaces are the level surfaces of a function

$\Sigma(x)$ this means $V = ad\Sigma$ for some a. Show that this implies $V \wedge dV = 0$. Show that the converse is also true.

[11.6] The 3-sphere S^3 can be represented as pairs of complex numbers (z, w) with $z\bar{z} + w\bar{w} = 1$. Write

$$z = r \cos(\theta/2) exp\frac{1}{2}i(\phi + \psi), \quad w = r \sin(\theta/2) \exp \frac{1}{2}i(\phi - \psi),$$

so that S^3 is $r = 1$. If C^2 has the metric

$$ds^2 = dzd\bar{z} + dwd\bar{w}$$

show that S^3 has the metric

$$ds^2 = (\theta^1)^2 + (\theta^2)^2 + (\theta^3)^2$$

where

$$\theta^1 + i\theta^2 = \frac{1}{2}e^{i\psi}(d\theta - i \sin\theta d\phi)$$

$$\theta^3 = \frac{1}{2}(d\psi + \cos\theta d\phi).$$

Calculate the curvature in this basis. This calculation reappears in chapter 26.

[11.7] In four-dimensional flat space-time let $f(x^a)$ be homogeneous of degree -4, i.e. $f(\lambda x^a) = \lambda^{-4}f(x^a)$. Show that the form $f\epsilon_{abcd}x^a dx^b \wedge dx^c \wedge dx^d$ is exact. How does this result generalize to a flat space of n dimensions?

[11.8] An ideal fluid with stress-energy tensor T^{ab} given by

$$T^{ab} = (\rho + p)u^a u^b - pg^{ab},$$

with $u_a u^a = 1$ and $\nabla_a T^{ab} = 0$, is said to be a *simple thermodynamical fluid* if there exist scalars S, T, and n such that

$$kTdS = dE + pdV \quad (k = \text{Boltzmann's constant})$$

and

$$d(n\epsilon_{abcd}u^a dx^b \wedge dx^c \wedge dx^d) = 0,$$

where $E = \rho/n$ and $V = 1/n$. S, T, n, E and V are then called the specific entropy, the temperature, the particle number density, the specific energy and the specific volume, respectively. Show that a necessary and sufficient condition for T^{ab} to characterize a simple thermodynamical fluid is that

$$(\dot{\rho}dp - \dot{p}d\rho) \wedge d\rho \wedge dp = 0,$$

where the dot denotes $u^a\nabla_a$. (ρ, p, u^a, and g_{ab} are all assumed to be smooth functions of the local coordinates x^a.) Can an ideal fluid admit more than one 'simple thermodynamic structure'?

12   The transition from Newtonian theory

The chief support of the theory is to be found less in that lent by observation hitherto than in its inherent logical consistency, in which it far transcends that of classical mechanics, and also in the fact that it solves the perplexing problem of gravitation and of the relativity of motion at one stroke in a manner highly satisfying to our reason.

—H. Weyl (**Space, Time, Matter**)

NOW WE CONFRONT the problem of constructing a theory of gravity that is consistent with the demands of special relativity and   that reduces to the Newtonian theory of gravity in the appropriate limit. The problem is to complete the following diagram:

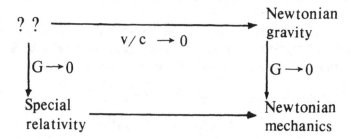

We cannot *deduce* what should go in the gap, but our guiding principle is the preservation of desirable features from Newtonian gravity and special relativity. With the help of hindsight we shall be led directly to Einstein's general relativity.

First, in special relativity we have a four-dimensional space-time, Minkowski space, with a metric η_{ab}, non-degenerate and with signature $(+ - - -)$. By use of η_{ab} we can define proper-time along any curve and distinguish, at each point, time-like, null and space-like vectors. The matter content of space-time is specified by a symmetric tensor T_{ab}, the energy momentum tensor, and the conservation of energy-momentum is guaranteed by the conservation equation

$$\frac{\partial T^{ab}}{\partial x^a} = 0. \tag{12.1}$$

In special relativity a particle subject to no forces moves on a time-like straight line.

In the new theory we wish to preserve as much as possible of this. However, the last statement is clearly in need of modification since in the presence of gravity a particle *cannot* be subject to *no* forces. The physical conception of gravity that we are trying to encode mathematically includes the idea that gravity acts on *everything*—every material particle in a gravitational field is subject to a gravitational force. Thus the system of interest is a particle subject to no force *except* gravity—a 'freely falling particle'.

According to Newtonian gravity a freely falling particle in a gravitational field with potential Φ moves according to the law of motion

$$ma_i = F_i = -m\nabla_i\Phi. \tag{12.2}$$

The first appearance of m here is the *inertial* mass of the particle defined as the appropriate constant of proportionality in Newton's second law of motion. The second appearance of m is of different origin: it is the 'gravitational mass' of the particle, which measures the response of the particle to a given gravitational field. A good analogy is with the charge of a charged particle, which is a measure of the response of the particle to a given electric field.

The fact that these two masses are proportional and can be cancelled from (12.2) must be regarded as an experimental fact as far as Newtonian theory is concerned. Cancelling the m's from (12.2), we obtain

$$\frac{d^2 x^i}{dt^2} = -\nabla_i\Phi \tag{12.3}$$

as the equation for the path of a freely falling particle. We may therefore interpret Newtonian gravity as saying that a gravitational field determines a preferred class of paths which sufficiently small objects (small enough that their own gravitational field does not interfere) must follow.

We retain this feature in the new theory by requiring that the new theory should have an *affine connection* that defines the preferred class of paths as its *geodesics*. Given the preferred class of paths, we must require of the completed theory that sufficiently small objects do follow these paths. This is referred to as the *geodesic hypothesis*. If the connection is to be related to ∇_i then the curvature, being the derivative of the connection, must be related to second derivatives of Φ. In particular, there must be some curvature.

We must also seek an analogue of the other ingredient of Newtonian gravity, namely the field equation

$$\nabla^2\Phi = 4\pi G\rho \tag{12.4}$$

relating the field to its source, the matter density ρ. The analogue of (12.4) must relate the curvature to the matter content specified by T_{ab}.

In summary, the new theory will be based on a four-dimensional manifold with a metric tensor g_{ab}, non-degenerate and with signature $(+ - - -)$, and an affine connection Γ^a_{bc}. The metric determines proper time as measured by physical clocks. We shall assume that the metric and connection are compatible in the sense that

$$\nabla_a g_{bc} = 0, \tag{12.5}$$

and for simplicity (by use of Occam's razor) we shall assume that the connection is torsion-free. By the *fundamental theorem of Riemannian geometry* we know that in this case the connection is uniquely determined by the metric. (It is possible, as noted earlier, to construct interesting theories in which the torsion is not zero. Then there is another field equation relating the torsion to a tensor field describing further properties of the material sources. The most important such theory is the so-called *Einstein-Cartan theory*.)

The matter content of the theory will be specified by a symmetric tensor T_{ab}. We shall assume that this is conserved with respect to the new connection in the sense that

$$\nabla_a T^{ab} = 0. \tag{12.6}$$

Since this equation involves the connection, it embodies an assumption about the interaction of matter and the gravitational field. We shall see that it is this assumption that justifies the geodesic hypothesis, i.e. it implies that sufficiently small objects do indeed follow geodesics.

Finally, we must seek a field equation of the form:

$$\{\text{tensor constructed from curvature}\}$$
$$= G \times \{\text{tensor constructed from matter}\}$$

that reduces to (12.4) in the Newtonian limit, i.e. in the limit of small velocities. In finding this equation, we shall be led by the geodesic hypothesis. The approach outlined here is to be a rough-and-ready, intuitive route to general relativity. It is not the only possible way, nor is it the route originally taken by Einstein. However, it will lead us in the chapters that follow to precise mathematical statements. The assumptions made here, completed by the field equations, will constitute the *general theory of relativity*.

Exercises for chapter 12

[12.1] Show that in Newtonian gravitation the equation of motion of a particle in a uniform gravitational field is the same as the equation of motion of a free particle referred to a (suitable) accelerated frame of reference.

[12.2] Suppose the energy-momentum tensor of a small particle introduced into a gravitational field is $T_{ab} = \rho u_a u_b$ where ρ is the density, and is zero outside a small world-tube, and u_a is the velocity, a time-like unit vector field on the world-tube tangent to the boundary. From the conservation equation (12.6) show that the vector field u^a is geodesic. What else does one learn from equation (12.6)?

[12.3] In a space of n dimensions define a tensor C_{abcd} by

$$C_{abcd} = R_{abcd} + \alpha(R_{ac}g_{bd} - R_{ad}g_{bc} + R_{bd}g_{ac} - R_{bc}g_{ad})$$
$$+ \beta R(g_{ac}g_{bd} - g_{ad}g_{bc})$$

in terms of the Riemann tensor, Ricci tensor and Ricci scalar and two constants α, β. Show that C_{abcd} has the same symmetries as the Riemann tensor. The constants α and β are chosen to make $C_{abc}{}^{b}$ vanish. If $n \geq 3$, what must they be? Show that if $n = 3$ then C_{abcd} vanishes identically. C_{abcd} is known as the *Weyl* tensor, or *conformal* tensor.

[12.4] Show that in the case $n = 4$ the Weyl tensor C_{abcd} satisfies $\nabla^{a}C_{abcd} = 0$ if $R_{ab} = 0$. What relation holds if $R_{ab} \neq 0$?

[12.5] Show that equation (12.3) can be written in the form of a geodesic equation in 4-dimensional space-time. Hence read off Γ^{a}_{bc} and calculate the associated 'curvature'. By looking at the symmetries of this 'Riemann tensor' show that it cannot be derived from a metric.

13   Einstein's field equations

LET US CONSIDER the relative motion of freely falling particles. In Newtonian gravitation first, suppose we have a family of paths $x^i(\sigma, \rho)$, $i = 1, 2, 3$ where ρ labels the paths and σ is time along the paths. The field of velocities is $V^i = \partial x^i/\partial \sigma$ and the vector $\eta^i = \partial x^i/\partial \rho$ defines a vector connecting infinitesimally neighbouring paths.

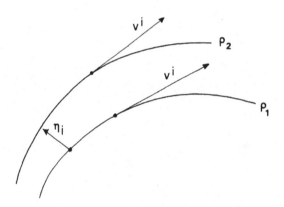

Figure 13.1. Relative motion in Newtonian gravitation.

If the gravitational potential is Φ then the equation of the paths is (12.3) or

$$\frac{\partial V^i}{\partial \sigma} = -\nabla_i \Phi. \tag{13.1}$$

Now if $\rho_2 = \rho_1 + \delta\rho$ where $\delta\rho$ is small, then we have

$$\left.\frac{\partial V^i}{\partial \sigma}\right|_{\rho_2} \simeq \left.\frac{\partial V^i}{\partial \sigma}\right|_{\rho_1} + \delta\rho \left.\frac{\partial^2 V^i}{\partial \sigma \partial \rho}\right|_{\rho_1}$$

$$= \left.\frac{\partial V^i}{\partial \sigma}\right|_{\rho_1} + \delta\rho \left.\frac{\partial^2 \eta^i}{\partial \sigma^2}\right|_{\rho_1}$$

and

$$\left.\frac{\partial V^i}{\partial \sigma}\right|_{\rho_1} = -\nabla_i \Phi|_{\rho_1}$$

$$\left.\frac{\partial V^i}{\partial \sigma}\right|_{\rho_2} = -\nabla_i \Phi|_{\rho_1} - \delta\rho\eta^j \nabla_i \nabla_j \Phi|_{\rho_1}.$$

Thus the relative acceleration of neighbouring paths is related to the *second* derivatives of Φ by the formula

$$\frac{\partial^2 \eta^i}{\partial \sigma^2} = -\eta^j \nabla_i \nabla_j \Phi. \tag{13.2}.$$

In a four-dimensional manifold M with curvature, the analogue of equation (13.2) is the *equation of geodesic deviation*. That is, if we suppose the paths are geodesics with respect to the metric connection of M then the relative acceleration is given by

$$\frac{D^2 \eta^a}{D\tau^2} = R_{bcd}{}^a V^b \eta^c V^d := \Phi^a_c \eta^c \tag{13.3}$$

where now all indices range over $0, 1, 2, 3$, and $R_{bcd}{}^a$ is the Riemann tensor of the metric connection. The role of the matrix of second derivatives of Φ is played by the tensor $\Phi_{ac} = R_{bcda} V^b V^d$. This is symmetric by virtue of the interchange symmetry on the Riemann tensor, and is orthogonal to the velocity V^a by virtue of the skew symmetries of the Riemann tensor:

$$\Phi_{ab} = \Phi_{ba}, \qquad \Phi_{ab} V^b = 0. \tag{13.4}$$

In Newtonian gravitation, the field equation (12.4) relates the trace of $\nabla_i \nabla_j \phi$ to the matter density ρ. Here the trace of Φ_{ab} is a component of the Ricci tensor:

$$\Phi^a_a = R_{bad}{}^a V^b V^d = R_{bd} V^b V^d, \tag{13.5}$$

which suggests that the field equations of general relativity should relate this to the matter density. Now, we know from special relativity that the matter density as measured by an observer with four-velocity V^a is $T_{ab} V^a V^b$ (cf. § 3.5–3.6) so we might try

$$R_{ab} V^a V^b \propto T_{ab} V^a V^b.$$

If this were to hold for all V^a we would be led to

$$R_{ab} = k T_{ab} \tag{13.6}$$

for some constant k to be determined by the Newtonian limit.

However, (13.6) cannot be *quite* right, since by assumption

$$\nabla_a T^{ab} = 0 \tag{13.7}$$

while from the contracted Bianchi identities (cf. § 7.3) we have

$$\nabla_a R^{ab} = \frac{1}{2} g^{ab} \nabla_a R. \tag{13.8}$$

We recall the definition of the Einstein tensor

$$G_{ab} = R_{ab} - \frac{1}{2}g_{ab}R$$

which by (13.8) *does* have vanishing divergence:

$$\nabla_a G^{ab} = 0.$$

Instead of (13.6) therefore we choose as field equations

$$G_{ab} = kT_{ab}. \tag{13.9}$$

It is a simple exercise in tensor algebra to see that this is equivalent to

$$R_{ab} = k(T_{ab} - \frac{1}{2}g_{ab}T) \tag{13.10}$$

where T is the trace

$$T = g^{ab}T_{ab}.$$

To fix k we recall from special relativity that

$$T = \rho + \Sigma_{i=1}^{3}p_i \tag{13.11}$$

where the p_i are the 'principal pressures'. Putting this in (13.10) we find

$$\Phi_a^a = R_{ab}V^aV^b = k(T_{ab}V^aV^b - \frac{1}{2}T) = \frac{1}{2}k(\rho - \Sigma p_i). \tag{13.12}$$

Now the Newtonian limit is the limit of small velocities compared to unity (i.e. compared to the speed of light). In particular, this means that the sum of the principal pressures should be much less than ρ (exercise: justify this on physical grounds). Thus we obtain agreement between (13.2) and (12.4) in this limit by taking $\frac{1}{2}k = -4\pi G$, i.e.

$$k = -8\pi G.$$

The field equations of general relativity motivated by comparison with Newtonian gravitation are therefore

$$R_{ab} - \frac{1}{2}g_{ab}R = -8\pi GT_{ab}, \tag{13.13}$$

or equivalently

$$R_{ab} = -8\pi G(T_{ab} - \frac{1}{2}g_{ab}T).$$

From this second form we see that in the *absence* of matter the field equations simply demand the *vanishing of the Ricci tensor*, from which we also deduce the vanishing of the Ricci scalar:

$$R_{ab} = 0. \tag{13.14}$$

These are referred to as Einstein's *vacuum field equations* and are the equations determining the free gravitational field in the space exterior to all masses.

One of the guiding features that has led us to (13.13) is the conservation equation (13.7). However, we now *postulate* (13.13) as a *basic assumption* of general relativity. Then (13.7) ceases to be an independent assumption and becomes a *consequence* of the field equations. By the same token, the consequences of (13.7), including notably the geodesic motion of small objects, become consequences of the field equations. Thus, in general relativity the equations of motion arise in a certain sense as a consequence of the field equations. This may be contrasted with Newtonian gravity, where the equation of motion (12.3) is quite independent of the field equation (12.4), and with electromagnetism, where the Lorentz force law for a charged particle in an electromagnetic field is independent of Maxwell's equations for the field itself.

Thus we see that Einstein's law of gravitation, in the absence of matter, is summarised in the brief and elegant equation $R_{ab} = 0$, whereas in the presence of gravitating matter we have his general field equations $G_{ab} = -8\pi G T_{ab}$. These relations are the goal and object of our study.

Exercises for chapter 13

[13.1] Use the equation of geodesic deviation to estimate some components of the Riemann tensor at the surface of the earth. (Hint: consider the relative acceleration of particles falling from rest either one above the other or side by side.)

[13.2] Consider a 'congruence' of time-like geodesics (i.e. a family of geodesics, one through each point of space-time), with unit tangent vector field t^a. Choose an orthogonal triad $e^a_{(i)}$, $i = 1, 2, 3$, of vectors orthogonal to t^a on each geodesic and parallelly propagate them along the geodesics. (i) Show that the metric tensor of space-time can be written

$$g_{ab} = t_a t_b - \delta_{ij} e_{a(i)} e_{b(j)}$$

where δ_{ij} is the Kronecker delta. Suppose that η^a is a connecting vector orthogonal to t^a along one geodesic γ, so that η^a is Lie-propagated along γ. Expand η^a in the triad as $\eta^a = \eta_i e^a_{(i)}$. (ii) Show that $D\eta_i = M_{ij}\eta_j$ where $M_{ij} = e^a_{(i)} e^b_{(j)} \nabla_a t_b$ and $D = t^a \nabla_a$. (iii) From the equation of geodesic deviation applied to η^a show that

$$D^2 \eta_i = \phi_{ij}\eta_j \text{ where } \phi_{ij} = R_{bcda} t^b t^d e^a_{(i)} e^c_{(j)}.$$

Deduce that $DM_{ij} + M_{ik}M_{kj} = \phi_{ij}$. (iv) Suppose that M_{ij} is decomposed into its anti-symmetric part A_{ij}, its symmetric trace-free part S_{ij} and its trace θ so that

$$M_{ij} = A_{ij} + S_{ij} + \frac{1}{3}\theta_{ij}.$$

Show that $\theta = -\nabla_a t^a$. (v) A small sphere of particles nearby at one instant is given by $\{\eta^i | \delta_{ij} \eta^i \eta^j = 1\}$. Show that at the next instant this sphere is rotated, expanded and sheared (i.e. made ellipsoidal) by amounts related to A_{ij}, θ, and S_{ij}, respectively. (vi) From parts (iii) and (iv) above show that

$$D\theta = S_{ij} S_{ij} - A_{ij} A_{ij} + \frac{1}{3}\theta^2 + \phi$$

where $\phi = \phi_{ii} = -R_{ab} t^a t^b$. Find similar propagation laws for A_{ij} and S_{ij}. Show that the propagation law for S_{ij} involves the Weyl tensor (exercise 12.3). The object of this exercise is to give a more explicit picture of the geometrical effects of curvature. Thus the Ricci curvature, which is tied directly to the matter content of space-time by the Einstein equations, generates *focussing* or *convergence* via θ, while the Weyl tensor, which constitutes the remainder of the curvature and may be thought of as the *free gravitational field*, generates shear via S_{ij}. The rotation A_{ij} is unaffected by curvature.

[13.3] Show that if the only energy in a region of space-time is electromagnetic then the curvature scalar vanishes there.

[13.4] Let T_{ab} be the electromagnetic energy tensor. (i) Show that if A^a and B^b are future null vectors then $T_{ab} A^a B^b \geq 0$. Deduce that if P^a and Q^b are future time-like vectors then $T_{ab} P^a Q^b \geq 0$. This condition on an energy tensor is called the *dominant energy condition* and is believed to hold not merely for electromagnetic but for *all* forms of matter. (ii) The physical meaning of (i) can be understood as follows. Let V^a be the 4-velocity of an observer and define $F^a = T_b^a V^b$. Show that F^a is the observer's Poynting 4-vector. It has component form {energy, Poynting 3-vector}. Now deduce from (i) that F^a is future time-like or null, and hence the velocity of energy flow cannot exceed the velocity of light.

[13.5] What does the dominant energy condition (cf. exercise 13.4) tell us about the Ricci curvature? Now look again at exercise 13.2 (vi) and observe, as a consequence, that for a rotation-free congruence ($A_{ij} = 0$) the convergence (θ) is subject to $D\theta > 0$. Deduce that once the geodesics start to converge then they are forced to focus. This phenomenon, as applied to light rays, is of great importance in understanding the collapse of stars and the nature of black holes.

[13.6] What is the magnitude of the scalar curvature associated with ordinary water? Is it legitimate to ignore the contribution of the water pressure in this calculation?

14 The slow-motion approximation

OUR DEVELOPMENT of general relativity has been assisted by a series of assumptions motivated by analogy with special relativity and Newtonian gravitation. We must now check that in the appropriate limit we can indeed recover Newtonian gravitation. This is the limit of small velocities, called the *slow motion approximation*. We assume that all velocities v involved are much less than the velocity of light (i.e. $v/c \ll 1$) and that the time derivatives of all quantities are much less than the space derivatives (in other words, the rate of change per year is much less than the rate of change per light year). Furthermore we assume that locally we can find a coordinate system that is in a suitable sense approximately Cartesian so that the metric tensor g_{ab} in this coordinate system is equal to the Minkowski metric η_{ab} plus a smaller term:

$$g_{ab} = \eta_{ab} + \varepsilon h_{ab}. \tag{14.1}$$

The contravariant metric g^{ab} in this approximation becomes:

$$g^{ab} = \eta^{ab} - \varepsilon h^{ab} \tag{14.2}$$

where the indices on h^{ab} are raised with η^{ab}, i.e.

$$\eta^{ab}\eta_{bc} = \delta^a_c, \quad h^{ab} = \eta^{ac}\eta^{bd}h_{cd}. \tag{14.3}$$

Also we are assuming that with respect to this coordinate system

$$\frac{\partial f}{\partial t} = O(\varepsilon) \times \frac{\partial f}{\partial x^i}, \quad i = 1, 2, 3 \tag{14.4}$$

for all functions of interest.

The aim is to relate the geodesic equation, with these assumptions, to the Newtonian force law (12.3), and relate the Einstein equations (13.13) to the Newton-Poisson equation (12.4). For the geodesic equation we need first the Christoffel symbols. These are

$$
\begin{aligned}
\Gamma^a_{bc} &= \frac{1}{2} g^{ad}(g_{bd,c} + g_{cd,b} - g_{bc,d}) \\
&= \frac{1}{2}\varepsilon \eta^{ad}(h_{bd,c} + h_{cd,b} - h_{bc,d}) + O(\varepsilon^2)
\end{aligned} \tag{14.5}
$$

by use of (14.1) and (14.2). Now the geodesic equations are

$$d^2 x^a/ds^2 + \Gamma^a_{bc}\frac{dx^b}{ds}\frac{dx^c}{ds} = 0, \tag{14.6}$$

but for a slowly moving particle with $v = O(\varepsilon)$ we have

$$\frac{dx^i}{ds} = \frac{dt}{ds}\frac{dx^i}{dt} = O(\varepsilon) \quad i = 1,2,3$$

$$\frac{dx^0}{ds} = \frac{dt}{ds} = 1 + O(\varepsilon).$$

Thus for a slowly moving particle the geodesic equations (14.6) reduce to

$$\frac{d^2x^i}{dt^2} + \Gamma^i_{00}\frac{dx^0}{ds}\frac{dx^0}{ds} = O(\varepsilon^2). \tag{14.7}$$

Now from (14.5) we have

$$\Gamma^i_{00} = -\frac{1}{2}\varepsilon(h_{0i,0} + h_{0i,0} - h_{00,i}) + O(\varepsilon^2)$$

$$= \frac{1}{2}\varepsilon h_{00,i} + O(\varepsilon^2) \qquad .$$

by use of (14.4) and $\eta^{ij} = -\delta^{ij}$. So at last the geodesic equation reduces to

$$\frac{d^2x^i}{dt^2} = -\frac{1}{2}\varepsilon\nabla_i h_{00}$$

and we obtain agreement with (12.3) if we identify

$$g_{00} = 1 + \varepsilon h_{00} = 1 + 2\Phi + O(\varepsilon^2) \tag{14.8}$$

where Φ is the Newtonian gravitational potential. For the Ricci tensor we have

$$R_{ab} = \Gamma^c_{ac,b} - \Gamma^c_{ab,c} + O(\varepsilon^2)$$

so that

$$R_{00} = -\Gamma^i_{00,i} + O(\varepsilon^2)$$

$$= -\frac{1}{2}\varepsilon h_{00,ii} + O(\varepsilon^2)$$

$$= -\nabla^2\Phi + O(\varepsilon^2). \tag{14.9}$$

In chapter 13 we saw that in the slow motion limit we have

$$-8\pi G(T_{00} - \frac{1}{2}g_{00}T) = -4\pi G\rho.$$

So by virtue of (14.9), Einstein's equations (13.13) reduce to the Newton-Poisson equation (12.4). In summary, the Newtonian force law *requires* us to make the identification (14.8). Then the Newtonian equation (12.4) follows from Einstein's equations.

It should be noted that *any* theory of gravitation that incorporates the geodesic hypothesis must make the identification (14.8) so that this is not a *unique* characteristic of general relativity; an experimental consequence of (14.8), such as we find in chapter 16, does not constitute a test of general relativity in the strict sense (except inasmuch as a negative result of such a test would falsify the theory).

Exercises for chapter 14

[14.1] If the assumptions of this section hold in a coordinate system (t, x^i) $i = 1, 2, 3$, show that they also hold in a transformed coordinate system

$$t' = t + \varepsilon f(x^i), \quad x'^i = x^i + \varepsilon g^i(x^j)$$

and that the only component of h_{ab} unchanged by this transformation is h_{00}. Thus the other components of h_{ab} have no invariant significance.

[14.2] (i) A general *gauge transformation* is $x^a \to x'^a = x^a + \varepsilon \zeta^a$ where ζ^a now depends on all four coordinates (cf. exercise 14.1). Under this transformation show that $h_{ab} \to h'_{ab} = h_{ab} - 2\zeta_{(a,b)}$. (ii) Rather than use h_{ab} directly it is often more useful to employ its trace-reversed form $\bar{h}_{ab}$ defined by:

$$\bar{h}_{ab} = h_{ab} - \frac{1}{2}\eta_{ab}h$$

where $h = h_c^c$. Show that we can make a gauge transformation which results in $\bar{h}_{ab}$ being divergence-free. This is the *de Donder gauge*.

[14.3] Show that the linearized form of Einstein's equations is

$$\varepsilon(\Box\,\bar{h}_{ab} + \eta_{ab}\bar{h}_{cd}{}^{,cd} - \bar{h}_{ac,}{}^c{}_b - \bar{h}_{bc,}{}^c{}_a) = 16\pi G T_{ab}$$

where $\bar{h}_{ab} = h_{ab} - \frac{1}{2}\eta_{ab}h$. In the first term $\Box$ is the usual flat-space D'Alembertian wave operator, and the three other terms serve merely to 'protect' the equation from gauge changes. Deduce that in empty space the choice of de Donder gauge (cf. exercise 14.2) reduces this to a simple wave equation:

$$\Box\,\bar{h}_{ab} = 0.$$

[14.4] A gravitational wave is a 'ripple' in the structure of empty space. When the wave is weak and planar we can represent it in the linear approximation by

$$\bar{h}_{ab} = Re[A_{ab}exp(ik_a x^a)]$$

where A_{ab} is a constant tensor. Show that k_a is null and orthogonal to A_{ab} (use the result of exercise 14.3). Such waves (which propagate at the speed of light) can excite particles into relative motion (geodesic deviation) and hence must be carrying energy. Einstein's theory predicts that gravitational waves *must* exist (e.g. as created by violent astrophysical events), and experimentalists are keenly engaged in attempts to detect them.

15   The Schwarzschild solution

THE FIELD EQUATIONS of general relativity are now at our disposal, and we know that general relativity reduces, in a suitable sense, to Newton's theory of gravitation in the slow motion limit. The next stage is to investigate, in a simple case, the new phenomena brought to light by general relativity. We shall construct the solution of the vacuum field equations corresponding to *the gravitational field outside an isolated spherically symmetric static body*. Then, thinking of this as the gravitational field of the sun (suitably approximated), we shall investigate properties of this solution by solving the geodesic equations. The solution in question here is the famous Schwarzschild solution.

A solution of the vacuum field equations is a pair consisting of a manifold M together with a metric g_{ab} on it satisfying (13.14). In a particular coordinate patch of M, (13.14) comprises ten non-linear partial differential equations for the ten unknown functions that are the components of the metric g_{ab} with respect to the coordinate basis. Clearly, the way to proceed is *not* to seek the general solution of these equations, any more than one would necessarily begin the study of electromagnetism by seeking the general solution of Maxwell's equations. (In fact, the general solution of the Einstein vacuum equations remains unknown, as of 1990; and if the aspiring student wishes to make a name for himself, therein lies an area of investigation.) Instead, we make some simplifying assumptions. Then we choose a form of the metric in a particular coordinate patch consistent with these assumptions, and depending therefore on a smaller number of functions. Finally, we equate the Ricci tensor of this simpler metric to zero and hope to get equations simple enough to be solvable.

This programme, if successful, will result in an explicit metric in one coordinate patch. For many purposes this is sufficient, but there remains the question of what is the underlying manifold M? We shall return to this point later. For now, the simplifying assumptions we wish to make on the metric are that it be *independent of time* and *spherically symmetric*. These assumptions may be given a precise mathematical expression in terms of *Killing vectors* but for the moment we shall proceed rather more intuitively.

Beginning with the time-independence, we assume that we can choose one coordinate $x^0 = t$ such that all metric components are independent of t:

$$\frac{\partial g_{ab}}{\partial t} = 0 \tag{15.1}$$

and further that the metric is unchanged if we change the sign of t. Then the metric must necessarily take the form

$$ds^2 = g_{ab}dx^a dx^b = g_{00}dt^2 + g_{ij}dx^i dx^j \quad (i,j = 1,2,3), \qquad (15.2)$$

where g_{00} and g_{ij} are functions of the spatial coordinates only. Next we wish to impose spherical symmetry. This means that on a surface of constant t we can find a radial coordinate R such that the surfaces of constant R are spheres. That is, they have the metric of a sphere whose radius is some function of R (independent of t by 15.1). If we introduce polar coordinates θ and ϕ on each sphere in the standard way, then the metric of one such sphere must take the form

$$-C^2(R)(d\theta^2 + \sin^2\theta d\phi^2).$$

for some function $C(R)$.

These spheres must be uniformly 'stacked' with increasing R so that we may suppose all the north poles and all the south poles to be lined up. This means we can line up every radius as a line of constant θ and ϕ and use the same angular coordinates on every sphere. The entire three-dimensional metric must therefore take the form

$$g_{ij}dx^i dx^j = -B^2(R)dR^2 - C^2(R)(d\theta^2 + \sin^2\theta d\phi^2).$$

There can be no $dRd\theta$ or $dRd\phi$ terms because the radial direction (direction of increasing R and fixed θ and ϕ) is orthogonal to the spheres of constant R.

Since g_{00} must also be a function only of R, say $A^2(R)$, the general metric consistent with our assumptions takes the form

$$ds^2 = A^2(R)dt^2 - B^2(R)dR^2 - C^2(R)(d\theta^2 + \sin^2\theta d\phi^2). \qquad (15.3)$$

One of the functions in (15.3) is clearly *redundant*, since we are always at liberty to define a new R-coordinate which is a function of the old. In particular, we define $r = C(R)$ and reduce (15.3) to

$$ds^2 = e^{2\lambda(r)}dt^2 - e^{2\nu(r)}dr^2 - r^2(d\theta^2 + \sin^2\theta d\phi^2). \qquad (15.4)$$

Here the form of g_{tt} and g_{rr} is chosen for later convenience. Thus the general spherically-symmetric time-independent metric depends on just two functions of the radial coordinate. If λ and ν are both zero then we recognize (15.4) as the metric of Minkowski space-time in spherical polar coordinates. Thus if (15.4) is to be the metric of an isolated system, we must demand that it approach the Minkowski metric at large distances, i.e. that λ and ν tend to zero as r tends towards infinity.

To impose the vacuum equations we must now calculate the Ricci tensor associated with this metric (see section 11.3). Labelling the coordinates $(t, r, \theta, \phi) = (x^0, x^1, x^2, x^3)$ we find

$$R_{00} = e^{2\lambda - 2\nu}\left(-\lambda'' + \lambda'\nu' - \frac{2\lambda'}{r} - \lambda'^2\right) \qquad (15.5)$$

$$R_{11} = \lambda'' + \lambda'^2 - \lambda'\nu' - \frac{2\nu'}{r} \qquad (15.6)$$

$$R_{22} = e^{-2\nu}(1 - e^{2\nu} + r(\lambda' - \nu')) \qquad (15.7)$$

$$R_{33} = R_{22}\sin^2\theta \qquad (15.8)$$

$$R_{ij} = 0 \quad \text{for } i \neq j.$$

Here the prime denotes differentiation with respect to r. We now equate these expressions to zero. This gives three equations on the two unknowns — but we are confident that they will be consistent, since otherwise there would be no spherically symmetric time-independent gravitational field!

From the first two we see at once that $\lambda' + \nu' = 0$, and therefore that $\lambda + \nu = $ constant. To satisfy the boundary condition, we take this constant to be zero. Now the vanishing of (15.7) gives

$$1 - e^{2\nu} - 2r\nu' = 0$$

which is equivalent to

$$(re^{-2\nu})' = 1,$$

so

$$e^{-2\nu} = 1 + \frac{k}{r} = e^{2\lambda}$$

for a constant of integration k. To identify k we see that

$$g_{00} = e^{2\lambda} = (1 + \frac{k}{r})$$

but from the discussion in chapter 14 and equation (14.8),

$$g_{00} \simeq (1 + 2\Phi + O(\epsilon^2)),$$

so if this metric is to represent the gravitational field of a spherical object of mass M we must have

$$\frac{k}{r} = 2\Phi = -\frac{2GM}{r}.$$

With this identification, (15.4) is the *Schwarzschild solution*:

$$ds^2 = (1 - \frac{2GM}{r})dt^2 - (1 - \frac{2GM}{r})^{-1}dr^2 - r^2(d\theta^2 + \sin^2\theta d\phi^2). \qquad (15.9)$$

In the next few chapters, we investigate various properties of the Schwarzschild solution. In particular, we find the orbits of planets around a spherical body by solving the geodesic equations, and we consider what happens at the 'critical radius' where the metric in the form (15.9) becomes singular. This is at

$$r = r_s = 2GM \ (= \frac{2GM}{c^2}) \qquad (15.10)$$

and is known as the *Schwarzschild radius*, where g_{tt} is zero and g_{rr} becomes infinite. For a spherical body with the mass of the sun, the Schwarzschild radius is about $3km$. More precisely, we have

$$G = 6.670 \times 10^{-8} \ cm^3 g^{-1}s^2,$$

approximately, for the gravitational constant, and

$$c = 2.988 \times 10^{10} \ cm \ s^{-1}$$

for the speed of light. Whence for the 'conversion' factor we obtain

$$G/c^2 = 7.421 \times 10^{-29} \ cm \ g^{-1}.$$

The mass of the sun is

$$M = 1.989 \times 10^{33} g.$$

So we have

$$GM/c^2 = 1.476 \times 10^5 \ cm,$$

and thus

$$r_s = 2GM/c^2 = 2.952 \ km.$$

To find the Schwarzschild radius associated with bodies of other masses one proceeds similarly.

Exercises for chapter 15

[15.1] A vector field k^a is said to be *hypersurface orthogonal* (HSO) if there exists a family of hypersurfaces such that k^a is proportional to the normal. Equivalently, k^a is HSO if there exist scalar functions U and V such that $k_a = U\nabla_a V$. Show that k_a is HSO $\implies k_{[a}\nabla_b k_{c]} = 0$. Now try to prove the converse. (Hint: the first stage is to show that, if $k_{[a}\nabla_b k_{c]}$ is zero and if X^a and Y^a are any two vector fields orthogonal to k^a then the commutator $X^b\nabla_b Y^a - Y^b\nabla_b X^a$ is also orthogonal to k^a.)

[15.2] A *Killing vector* k^a is a vector field for which

$$\mathcal{L}_k g_{ab} := k^c\nabla_c g_{ab} + g_{ac}\nabla_b k^c + g_{bc}\nabla_a k^c = 0$$

Show that this expression is independent of (symmetric) connection. If k^a is a Killing vector and we choose coordinates (x^0, x^1, x^2, x^3) such that $k^a\nabla_a = \partial/\partial x^0$ show that $\partial g_{ab}/\partial x^0 = 0$. If k^a is also HSO show that

$$\frac{\partial}{\partial x^i}\left(\frac{1}{g_{00}}g_{0j}\right) = \frac{\partial}{\partial x^j}\left(\frac{1}{g_{00}}g_{0i}\right) \quad i = 1, 2, 3.$$

Deduce that $g_{0i} = g_{00}\partial f/\partial x^i$. for some function $f(x^i)$. Now make the coordinate transformation

$$x'^0 = x^0 - f(x^i), \quad x'^i = x^i.$$

Show that in this coordinate system $g_{0i} = 0$. Thus the metric form (15.2) is equivalent to the assumption of a hypersurface orthogonal Killing vector.

[15.3] From the Killing vector equation in the form

$$\nabla_a k_b + \nabla_b k_a = 0$$

deduce that

$$\nabla_a \nabla_b k_c = R_{bcad} k^d.$$

This relates second derivatives of k_a algebraically to k_a. Clearly, by differentiating this relation, all higher derivatives of k_a can be related to lower derivatives. What is the maximum number of Killing vectors that an n-dimensional space can have?

[15.4] If k^a is a Killing vector and ξ^a is the tangent to an affinely parametrised geodesic γ, show that $g_{ab}k^a\xi^b$ is constant along γ. What is the converse of this? Is it true?

[15.5] If k^a and ℓ^a are Killing vectors, show that any constant linear combination of them is a Killing vector, as is their commutator $m^a = k^b\nabla_b\ell^a - \ell^b\nabla_b k^a$.

[15.6] Use exercise [8.2] to obtain a neater derivation of the fact that the commutator of two Killing vectors is a Killing vector.

[15.7] A *rotating* star's geometry is no longer spherically symmetric. However, it still has two symmetries, described by the Killing vectors T^a (time translation) and Φ^a (axial rotation). Assuming there are no other independent Killing vectors, show that T^a and Φ^a commute. (Hint: consider the geometry far from the star.)

16   Gravitational red-shift and time dilation

A REMARKABLE FEATURE of general relativity that can conveniently be introduced at this point is the *gravitational red-shift*.

The physical notion underlying the gravitational red-shift is that light *loses energy* in climbing out of a gravitational potential well. Thus light of a particular frequency emitted, say, at the surface of the sun, where the gravitational potential is large and negative, should be shifted towards the red, i.e. should have a *lower* frequency, when observed at the earth where the gravitational potential is less negative.

The geometrical picture for this process is familiar from special relativity (see section 3.2). We suppose the emitter has world line E with unit four-velocity u^a and the observer has world line O with four-velocity v^a. If the light ray is an affinely parametrized null geodesic L with tangent vector ℓ^a then the emitted frequency ν_E and observed frequency ν_O are just

$$\nu_E = g_{ab}u^a\ell^b, \quad \nu_O = g_{ab}v^a\ell^b. \tag{16.1}$$

In special relativity, we know that these may differ due to Doppler shifting if E and O are in relative motion. In general relativity, they will differ even if E and O are both 'at rest' in the Schwarzschild solution. By 'at rest' here we shall mean that the velocities of E and O are both parallel to the time-like Killing vector k^a of the Schwarzschild solution (see exercise 15.2).

If the world-line E is at radius r_E and the world line O at radius r_O then

$$u^a = k^a(v(r_E))^{-\frac{1}{2}}, \quad v^a = k^a(v(r_O))^{-\frac{1}{2}} \tag{16.2}$$

where

$$v(r) = g_{ab}k^ak^b = 1 - \frac{2GM}{r}.$$

The last step uses (15.9) and the fact that in these coordinates $k^a = \delta_0^a$.

To calculate the frequencies (16.1) we also need to know $g_{ab}k^a\ell^b$ at E and O, but since L is affinely parametrized and k^a is a Killing vector, $g_{ab}k^a\ell^b$ is *constant* along L (exercise 15.4)! Thus the ratio of observed to emitted frequencies is

$$\frac{\nu_0}{\nu_E} = \frac{g_{ab}k^a\ell^b}{(v(r_O))^{\frac{1}{2}}} \frac{(v(r_E))^{\frac{1}{2}}}{g_{ab}k^a\ell^b} = (1 - \frac{2GM}{r_E})^{\frac{1}{2}}(1 - \frac{2GM}{r_O})^{-\frac{1}{2}}. \tag{16.3}$$

If r_E and r_O are both large compared to r_s then we have the following approximate expression for the red-shift:

$$\frac{\Delta\nu}{\nu_E} = \frac{\nu_O - \nu_E}{\nu_E} \simeq -\frac{GM}{r_O} + \frac{GM}{r_E} = \Delta\Phi, \tag{16.4}$$

The expression (16.3) also tells us that there is a *time dilation* between E and O. In a small interval δS_E of proper time at E, n wave crests will be emitted where $n = \nu_E \delta S_E$. These n wave crests will arrive at O in an interval δS_O where $n = \nu_0 \delta S_O$. Thus by (16.3) we have

$$\delta S_O = (1 - \frac{2GM}{r_O})^{1/2}(1 - \frac{2GM}{r_E})^{-1/2}\delta S_E. \qquad (16.5)$$

This is a time dilation due to the gravitational field: the clock at the smaller value of r, i.e. at the smaller (more negative) gravitational potential runs *slower*.

These effects are indeed observed in nature; but this is not regarded as a *direct* test of general relativity because (16.4) follows from the identification (14.8) which, as we have remarked, would be made in almost any theory with the same mathematical framework as general relativity. However, since we know general relativity now to be the *correct* classical description of gravitation, the gravitational red-shift can be embraced as one of its significant consequences.

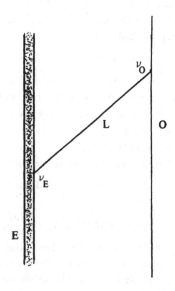

Figure 16.1. The gravitational red-shift: light emitted at frequency ν_E from a massive body E will be measured at reception by an observer O to have a *lower* frequency ν_0.

Exercises for chapter 16

[16.1] In the course of a year standing on the earth, how much more or less does one's head age than one's feet?

[16.2] The energy of a photon is $h\nu$, where h is Planck's constant $(6.626 \times 10^{-27} erg\, s)$. Use Newtonian conservation of energy to rederive (16.4).

[16.3] A small spaceship has constant acceleration (g) as it travels through a region of space far from any stars. Light is fired from the rear to the front. Calculate the Doppler shift. The ship now lands on a planet called the Earth and the experiment is repeated. Einstein's 'equivalence principle' (cf. exercise 12.1) tells us that the results of both experiments must be identical. Use this to deduce (16.4) yet again.

[16.4] In the text the number of waves emitted and received was used to deduce $\nu_O \delta S_O = \nu_E \delta S_E$ and hence, by (16.1), $\ell_a u^a \delta S_E = \ell_a v^a \delta S_O$. Here is another way. Let Θ be the phase of the waves so that $\Theta = constant$ are the wavefronts and $\ell_a = \nabla_a \Theta$. Check that Θ doesn't change along the rays. Consider two wavefronts differing in phase by $\delta\Theta$. How much time elapses for the observer in travelling from one surface to the next? Deduce the result.

[16.5] The sun has mass 1.989×10^{33} g and radius 6.96×10^{10} cm. The radius of the Earth's orbit is 1.49×10^{13} cm. Calculate the gravitational red-shift experienced by a photon travelling from the Sun's surface to the Earth, ignoring the Earth's mass and its rotation about the sun. The Earth's mass is 5.98×10^{27} g and its mean radius is 6.37×10^{8} cm. How does this effect the result of the previous calculation? And what about the transverse Doppler effect induced by the Earth's rotation about the Sun? (1 *year* $\sim 3.16 \times 10^7$ s.) The sun itself rotates once in 2.14×10^6 seconds—does this affect the result? (See section 3.4.) What about the Earth's rotation about its own axis? (1 *day* $\sim 8.64 \times 10^4$ s.)

17   The geodesic equation for the
Schwarzschild solution

LET US BEGIN by reviewing some of what we know about geodesics from chapter nine. A curve γ with tangent vector T^a is a *geodesic* if

$$T^b \nabla_b T^a = k T^a \tag{17.1}$$

for some function k along γ. If γ is given parametrically as $x^a(\lambda)$ in some coordinate patch then with $T^a = dx^a/d\lambda$ (17.1) becomes

$$\frac{d^2 x^a}{d\lambda^2} + \Gamma^a_{bc} \frac{dx^b}{d\lambda} \frac{dx^c}{d\lambda} = k \frac{dx^a}{d\lambda}. \tag{17.2}$$

Under a reparametrization $\lambda \to \lambda' = f(\lambda)$ the form of (17.2) is unchanged but with

$$k \to k' = k \left(\frac{d\lambda}{d\lambda'} \right)^2 + \frac{d^2\lambda}{d\lambda'^2}. \tag{17.3}$$

In particular, we may choose the new parameter such that k' vanishes. The geodesic is then said to be *affinely parametrized* and the new parameter is an *affine parameter*. Clearly from (17.3) any two affine parameters, say λ', λ'', are related by an 'affine transformation': $\lambda'' = a\lambda' + b$ where a and b are constants. Now for an affinely parametrized geodesic (17.1) and (17.2) reduce to

$$T^b \nabla_b T^a = 0 \tag{17.4}$$

$$\frac{d^2 x^a}{d\lambda^2} + \Gamma^a_{bc} \frac{dx^b}{d\lambda} \frac{dx^c}{d\lambda} = 0. \tag{17.5}$$

This means that $g_{ab} T^a T^b$ is *constant* along γ. If γ is a *null* geodesic of course this is true for any parametrization and the constant is zero. However for a time-like geodesic we may use an *affine transformation* to make this constant equal to one. The special affine parameter obtained in this way is *proper time*, and is defined up to changes of origin, i.e. additive constants only. (Similarly, on a space-like geodesic the constant can be set equal to -1.)

It would appear from (17.5) that to write down the geodesic equation for a given metric we have first to calculate the forty Christoffel symbols of that metric. Fortunately, however, we shall see that this is not the case!

Intuitively a geodesic joining two points p and q is expected to be the 'shortest path' from p to q. Accordingly, we suppose first of all that p and q have at least one time-like path joining them and we consider all parametrized time-like or null curves $x^a(\lambda)$ with $\lambda = \lambda_0$ at p and $\lambda = \lambda_1$ at q. The length from p to q is then

$$S = \int_{\lambda_0}^{\lambda_1} L d\lambda \tag{17.6}$$

where $L(x^a, \dot{x}^a) = (g_{ab}\dot{x}^a\dot{x}^b)^{\frac{1}{2}}$ and $\dot{x}^a = dx^a/d\lambda$. We may find the curve that gives a stationary value to S from the Euler-Lagrange equation applied to L. This is

$$\frac{d}{d\lambda}\left(\frac{\partial L}{\partial \dot{x}^a}\right) - \frac{\partial L}{\partial x^a} = 0$$

where we use a dot for $d/d\lambda$, which reduces to

$$\ddot{x}^a + g^{ae}g_{eb,c}\dot{x}^b\dot{x}^c - \frac{1}{2}g^{ae}g_{bc,e}\dot{x}^b\dot{x}^c = \frac{1}{2}\dot{x}^a(g_{bc,d}\dot{x}^b\dot{x}^c\dot{x}^d), \tag{17.7}$$

and this, as expected, is just (17.2), the geodesic equation with arbitrary parametrization. However we can do better than this. If we consider instead (17.6) but with

$$L = g_{ab}\dot{x}^a\dot{x}^b \tag{17.8}$$

then the Euler-Lagrange equation is (17.5), the geodesic equation with affine parametrization. Not only do we not need the Christoffel symbols before we can write down the geodesic equation, we can use the geodesic equation to find the Christoffel symbols! So much the better! .

We illustrate the procedure by considering the Schwarzschild solution. In this case the Lagrangian is given by

$$L = (1 - \frac{2M}{r})\dot{t}^2 - (1 - \frac{2M}{r})^{-1}\dot{r}^2 - r^2(\dot{\theta}^2 + \sin^2\theta\dot{\phi}^2). \tag{17.9}$$

We have simplified the notation here by omitting G or equivalently by choosing units such that $G = 1$. As with the speed of light c it is always possible by a consideration of dimensions to reintroduce the correct powers of G into any given expression of interest. The Euler-Lagrange equations can be calculated to be:

$$\left(\frac{\partial L}{\partial \dot{t}}\right)^{\bullet} - \frac{\partial L}{\partial t} = (2(1 - \frac{2M}{r})\dot{t})^{\bullet} = 0 \tag{17.10a}$$

$$\left(\frac{\partial L}{\partial \dot{r}}\right)^{\bullet} - \frac{\partial L}{\partial r} = (-2(1 - \frac{2M}{r})^{-1}\dot{r})^{\bullet} - \frac{2M}{r^2}(1 - \frac{2M}{r})^{-2}\dot{r}^2$$
$$+ 2r(\dot{\theta}^2 + \sin^2\theta\dot{\phi}^2) - \frac{2M}{r^2}\dot{t}^2 \tag{17.10b}$$

$$\left(\frac{\partial L}{\partial \dot{\theta}}\right)^{\bullet} - \frac{\partial L}{\partial \theta} = (-2r^2\dot{\theta})^{\bullet} + 2r^2\sin\theta\cos\theta\dot{\phi}^2 = 0 \tag{17.10c}$$

$$\left(\frac{\partial L}{\partial \dot{\phi}}\right)^{\bullet} - \frac{\partial L}{\partial \phi} = (-2r^2\sin^2\theta\dot{\phi})^{\bullet} = 0 \tag{17.10d}$$

and these are therefore the equations for t, r, θ, ϕ as functions of affine parameter λ. If desired we can read off the Christoffel symbols from (17.10). For example from (17.10a) in the form

$$\ddot{t} + \frac{2M}{r^2}(1 - \frac{2M}{r})^{-1}\dot{t}\dot{r} = 0$$

we deduce, by comparing this relation with (17.5), that

$$\Gamma^t_{tr} = \frac{M}{r^2}(1 - \frac{2M}{r})^{-1}, \qquad \Gamma^t_{ab} = 0 \quad \text{otherwise.}$$

When it comes to solving the geodesic equations we are still faced with four second-order non-linear ordinary differential equations. We have one 'first integral', i.e. an expression relating only first derivatives to a constant of integration, immediately. This is L, which is zero or ± 1 for null, time-like or space-like geodesics, respectively. To solve the geodesic equation, we would like as many first integrals as there are unknowns, since we would then have only first-order differential equations.

We saw in exercise [15.4] that another way of obtaining first integrals is by use of Killing vectors. If K^a is a Killing vector and T^a is tangent to an affinely parametrized geodesic γ then

$$Q = g_{ab}T^a K^b \tag{17.11}$$

is constant along γ. Now (17.11) equates a linear expression in $T^a = \dot{x}^a$ to a constant and so is a first integral. We might therefore hope to find more first integrals of (17.10) by finding a set of Killing vectors associated with the Schwarzschild solution.

Again, however, it is the converse process that is easier. There are two *obvious* first integrals of (17.10), namely

$$E = (1 - \frac{2M}{r})\dot{t} \tag{17.12}$$

and

$$J = r^2 \sin^2\theta\,\dot{\phi} \tag{17.13}$$

since (17.10a) is just $\dot{E} = 0$ and (17.10d) is $\dot{J} = 0$. In the language of classical mechanics, (17.12) and (17.13) correspond to the 'ignorable coordinates' in the Lagrangian (17.9). We can therefore actually use (17.11) in reverse. Since $E = g_{ab}T^a K_1^b$ for

$$K_1^b = (1,0,0,0), \tag{17.14}$$

and $J = g_{ab}T^a K_2^b$ for

$$K_2^b = (0,0,0,1), \tag{17.15}$$

and since E and J are constant along geodesics we deduce that K_1^a and K_2^a are Killing vectors of the Schwarzschild solution. In fact, K_1^a has already been exploited in chapter 16. We shall find some more Killing vectors in exercise [17.3].

We now have three first integrals, L, E and J. While it is possible to continue and find more, we can argue that these are sufficient. This is because any geodesic of the Schwarzschild solution must, by the spherical symmetry, remain in a plane going through the centre. To prove this, it is enough to check that a geodesic from a point in the equatorial plane initially tangent to the equatorial plane will be confined to the equatorial plane, i.e. if initially $\theta = \pi/2$ and $\dot{\theta} = 0$ then θ is always $\pi/2$. For this we consider (17.10c), the geodesic equation that which determines $\ddot{\theta}$. We have

$$\ddot{\theta} - 2\frac{\dot{r}}{r}\dot{\theta}^2 - \sin\theta\cos\theta\dot{\phi}^2 = 0$$

so that indeed if $\sin\theta = \pi/2$ and $\dot{\theta} = 0$ then $\ddot{\theta} = 0$ and θ will be always $\pi/2$.

From now on, we need only consider geodesics in the equatorial plane so our three constants are sufficient. The geodesic equations reduce to three first-order equations: (17.12), (17.13), and the equation for the constant L, all restricted to $\theta = \pi/2$.

Exercises for chapter 17

[17.1] Check the assertions made in the text, viz. that (17.7) is (17.2), that $g_{ab}T^aT^b$ is constant along affinely parametrized geodesics, and that (17.8) leads to (17.5).

[17.2] Would you expect a geodesic always to be the shortest path between two points? (Hint: how long is a null curve?)

[17.3] Show that the two quantities

$$P = r^2(\cos\phi\dot{\theta} - \sin\theta\cos\theta\cos\phi\dot{\phi})$$

$$Q = r^2(\sin\phi\dot{\theta} + \sin\theta\cos\theta\cos\phi\dot{\phi})$$

are first integrals of the geodesic equation for the Schwarzschild solution. Deduce two more Killing vectors, say K_3^a and K_4^a. What are the commutators between K_2^a, K_3^a and K_4^a? Notice that $Q\sin\theta\cos\phi - P\sin\theta\sin\phi - J\cos\theta = 0$. Of what surface is this the equation?

[17.4] A particle in equatorial orbit is perturbed out of its plane by $\delta\theta$. Show that $\delta\theta$ oscillates about zero, and hence that the orbit is *stable*.

[17.5] (i) In Newtonian theory what is the total energy (including its mass) of a particle of mass m at rest at distance r from a spherical star of mass M? (ii) In the Schwarzschild geometry we can take 'at rest' to mean that T^a is proportional to Killing vector K_1^a. By evaluating $E = T_aK_1^a$ for a particle at rest far from the star and comparing the result with (i), deduce that E represents energy per unit mass.

[17.6] (i) Show that in Newtonian theory the gravitational field is zero inside an empty spherical shell of matter. (ii) In chapter 15 it was established that spherical symmetry leads to the Schwarzschild geometry. Use this fact to show that the result of (i) is also true in Einstein's theory.

[17.7] A real star will be slightly *oblate* rather than perfectly spherical. Show that in Newtonian theory this means that the potential in the equatorial plane will have a second term proportional to r^{-3}.

[17.8] The equation of motion of a charged particle under the influence of an electromagnetic field F_{ab} given by the Lorentz equation

$$mu^a \nabla_a u^b = qF^b{}_c u^c$$

where m is the particle's mass and q is its charge. The particle velocity vector field u^a is normalized so that $u^a u_a = 1$. Suppose that ξ^a is a Killing vector and that the Lie derivative of F_{ab} with respect to ξ^a vanishes: $\mathcal{L}_\xi F_{ab} = 0$. Show that $F_{ab}\xi^b = \nabla_a \phi$ for some scalar field ϕ, and hence that $I = mu^a \xi_a + q\phi$ is a constant of motion for the charged particle: $u^a \nabla_a I = 0$. (Assume that m and q are constant.)

[17.9] Find a Lagrangian that leads to the Lorentz equation. Hint: $F_{ab} = \nabla_{[a} A_{b]}$.

[17.10] The exterior gravitational field of a spherical body is described by the metric

$$ds^2 = (1 - \frac{2M}{r})dt^2 - (1 - \frac{2M}{r})^{-1}dr^2 - r^2(d\theta^2 + \sin^2\theta d\phi^2).$$

For time-like geodesics in the equational plane show that the quantities J and E defined by

$$J = r^2 \frac{d\phi}{ds}, \quad E = (1 - \frac{2M}{r})\frac{dt}{ds}$$

are constants of the motion, where s is the proper time. Show, for such orbits, that the equations on r can be expressed in the form

$$(\frac{dr}{ds})^2 + (1 + \frac{J^2}{r^2})(1 - \frac{2M}{r}) = E^2$$

$$\frac{d^2r}{ds^2} + \frac{M}{r^2} - \frac{J^2}{r^3} + 3\frac{J^2 M}{r^4} = 0.$$

For a circular orbit at radius $r = R$ show that

$$R^3(\frac{d\Phi}{dt})^2 = M, \quad J^2 = MR(1 - \frac{3M}{R})^{-1}.$$

Show, by setting $r = R + \epsilon\xi(s)$ with ϵ small, that a circular orbit is stable if and only if $R > 6M$.

18 Classical tests

One of the most interesting problems of the astronomer at present is whether the motions of the heavenly bodies, as determined by our most refined methods of observation, go on in rigorous accordance with the law of gravitation. ... In 1845 Le Verrier found that the centennial motion of the perihelion of Mercury derived from observation was greater by 35″ than it should be from the gravitation of other planets, and his result has been more than confirmed by subsequent investigations, the most recent discussion of observations showing the excess of motion to be 43″ per century. In this case there can be no doubt as to the correctness of the theoretical result, since the computation of the secular motion of the perihelion is a comparatively simple process. It follows that either Mercury must be acted upon by some unknown body or the theory of gravitation needs modification.

—Simon Newcomb (**Encyclopaedia Britannica**, tenth edition, 1902)

WE PROCEED to solve the geodesic equations in the Schwarzschild solution and use the solution to describe the classical   tests of general relativity. These are the *precession of the perihelion of planetary orbits* and the *bending of light by the sun*, effects that arise from the small differences between orbits in Newtonian gravitation and orbits, i.e. geodesics, in general relativity.

For time-like geodesics we have the once-integrated geodesic equations in the form:

$$J = r^2 \dot{\phi}, \tag{18.1}$$

$$E = (1 - \frac{2M}{r})\dot{t}, \tag{18.2}$$

$$L = (1 - \frac{2M}{r})\dot{t}^2 - (1 - \frac{2M}{r})^{-1}\dot{r}^2 - r^2 \dot{\phi}^2 = 1, \tag{18.3}$$

where J, E, and L are constants of the motion.

We solve (18.3) for $\dot{r}$, eliminating $\dot{\phi}$ and $\dot{t}$ with the aid of (18.1) and (18.2):

$$\dot{r}^2 = (E^2 - 1) + \frac{2M}{r} - \frac{J^2}{r^2} + \frac{2MJ^2}{r^3}. \tag{18.4}$$

We now introduce a new variable $u = r^{-1}$, following the device commonly used in Newtonian theory. What we seek is the equation of the orbit, that is u as a function of ϕ. From (18.4) and (18.1) we find

$$(\frac{\partial u}{\partial \phi})^2 = \frac{\dot{u}^2}{\dot{\phi}^2} = \frac{\dot{t}^2}{r^4 \dot{\phi}^2} = \frac{1}{J^2} f(u) \tag{18.5}$$

where

$$f(u) = (E^2 - 1) + 2Mu - J^2 u^2 + 2MJ^2 u^3. \tag{18.6}$$

We may now integrate (18.5) to find, if $u = u_0$ at $\phi = \phi_O$:

$$\int_{u_0}^{u} \frac{du'}{\sqrt{f(u')}} = \int_{\phi_0}^{\phi} d\phi', \tag{18.7}$$

and in principle this gives the orbit equation explicity. Unfortunately, since $f(u)$ is a cubic polynomial, (18.7) cannot be integrated in terms of elementary functions. (The enterprizing student may wish to attempt the integration by use of elliptic functions.) Instead, we investigate the *qualitative* behaviour of the orbits, and then find an approximate solution. Clearly the nature of the orbit depends on the positions of the zeros of the polynomial $f(u)$, which in turn depend on the values of the constants E and J. We are interested in the coordinate range $\infty > r \geq 2M$ or $0 < u \leq \frac{1}{2M}$. We find

$$f(\frac{1}{2M}) = E^2 > 0,$$
$$f(0) = E^2 - 1,$$
$$\dot{f}(0) = 2M > 0,$$
$$\dot{f}(\frac{1}{2M}) = 2M + \frac{J^2}{M} > 0,$$

Since $f(u)$ is a cubic, several distinct cases arise, and the various possibilities are as shown in the figure below:

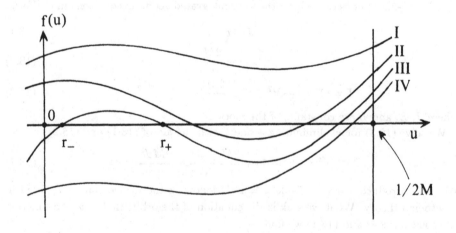

Figure 18.1 The graph of $f(u)$ showing the four different cases.

Case I: $f(u)$ is positive throughout. The orbit has no turning points, so the orbiting body plunges into the central body.

Case II: $f(u)$ is positive from $u = 0$ ($r = \infty$) down to a certain value of u, so u is monotonic in ϕ to here. Then there is a turning point. This is a *hyperbolic* orbit—the orbiting body comes in, reaches a closest approach and goes back out. There is also a region near $u = \frac{1}{2M}$ where $f(u)$ is positive. This corresponds to the orbits of bodies rising from the central body and falling back.

Case III: Here $f(u)$ is only positive in a region of u not including zero, that is for $r_- \leq r \leq r_+$ for some finite $r_\pm$. The orbiting body is confined between these values of r which are therefore perihelion and aphelion. This is an 'elliptic' orbit. Note that $E^2 < 1$ for this case. Again there are orbits arising from the central body.

Case IV: Here there are only orbits arising from the central body.

The terms 'elliptic' and 'hyperbolic' are only conventional here and refer to the fact that the analogous orbits in the Newtonian case actually are ellipses and hyperbolae. To see this and to prepare for the approximate solution to (18.5) we recall the Newtonian calculation (exercise 18.1). There we have

$$\left(\frac{\partial u}{\partial \phi}\right)^2 = \frac{1}{J^2}(h + 2Mu - J^2u^2) \tag{18.8}$$

and the solution is

$$u = \frac{M}{J^2}(1 + e\cos\phi) \tag{18.9}$$

where

$$e^2 = 1 + h\frac{J^2}{M^2}$$

and the arbitrary constant has been chosen so that $du/d\phi = 0$ at $\phi = 0$. This orbit is then an ellipse or hyperbola with focus at the origin according as to whether e^2 is less than or greater than one, i.e. according as to whether h is positive or negative.

Comparing (18.8) with (18.5) and (18.6) and identifying h with $E^2 - 1$, we see that the only difference is the cubic term in u. Also, from the figure, we have qualitative agreement in that orbits are elliptic or hyperbolic depending on the sign of $E^2 - 1$. For the orbit of the earth around the sun according to Newtonian gravitation we find, for various terms in $f(u)$, that (cf. the data noted in exercise 16.5):

$$2Mu \sim 10^{-8}, \quad J^2u^2 \sim 10^{-8};$$

and thus

$$2MJ^2u^3 \sim 10^{-16};$$

and, as expected, the cubic term can be treated as small. Therefore we seek a solution to (18.5) of the form of (18.9) plus a perturbation:

$$u = \frac{M}{J^2}(1 + e\cos\phi) + V. \tag{18.10}$$

Substituting this in (18.5) and keeping only the first order terms in V we find

$$\sin\phi\frac{\partial V}{\partial \phi} = V\cos\phi - \frac{M^3}{eJ^4}(1 + e\cos\phi)^3,$$

whence

$$V = \frac{M^3}{J^4}[\frac{(1+3e^2)}{e}\cos\phi + 3(1 + \frac{e^2}{2}) - \frac{e^2}{2}\cos 2\phi + 3e\phi\sin\phi]. \qquad (18.11)$$

We may analyse this expression as follows: the first and third terms are periodic—that is, they may change the shape of the orbit slightly but the orbit will remain closed with a period of 2π in ϕ. Likewise, the second term just adds a constant which is also periodic, but the *fourth* term is *not*. As ϕ increases from 0 to 2π, this term is not periodic and will steadily mount up. If we think of the period of the orbit as the angle in ϕ between two successive perihelia then we must look at the turning points to see what this is. We have

$$\frac{\partial u}{\partial \phi} = -\frac{Me}{J^2}\sin\phi + \frac{M^3}{J^4}[-\frac{1}{e}\sin\phi + e^2\sin 2\phi + 3e\phi\cos\phi]. \qquad (18.12)$$

This is zero at $\phi = 0$, and then the next zero is not at $\phi = \pi$, as it would be in Newtonian gravitation—but at, say, $\phi = \pi + \varepsilon$. Substituting into (18.12) we find

$$\varepsilon = \frac{3M^2}{J^2}\pi.$$

Thus in one complete revolution, the perihelion moves on from where it 'would be' in Newtonian theory by an amount

$$2\varepsilon = \frac{6M^2\pi}{J^2} = \frac{6M\pi}{L} \quad \text{radians} \qquad (18.13)$$

where L is the semi-latus-rectum of the Newtonian orbit,

$$L = \frac{1}{2}(1 - e^2) \times \text{ major axis}$$
$$= \frac{J^2}{M}.$$

This is the celebrated 'perihelion advance' predicted by general relativity. Although there are other changes in the orbit, the others are all periodic and, for example, for the planets orbiting the sun, could be accounted for by departures from exact spherical symmetry. This is not the case for the perihelion advance.

To calculate the perihelion advance in particular cases, we need to restore G and c to (18.13). Inserting the appropriate powers gives

$$2\varepsilon = 6\frac{GM}{c^2L}\pi \text{ radians}$$

with

$$L = \frac{J^2}{GM}.$$

For planets orbiting the sun we must take for M the mass of the sun $M = 1.989 \times 10^{33}$ g together with $G = 6.670 \times 10^{-8}$ $cm^3 g^{-1} s^{-2}$ and $c = 2.998 \times 10^{10}$ cm s^{-1} to find the Schwarzschild radius of the sun

$$r_s = \frac{2GM}{c^2} = 2.952 \times 10^5 \ cm,$$

as given in chapter 15. The advance is *largest* when L is as *small* as possible, namely for the planet Mercury, for which $L = 55.5 \times 10^{11}$ cm. With these numbers we find by use of (18.13) that

$$2\varepsilon = 5.03 \times 10^{-7} \text{ radians.}$$

This microscopic quantity is *the perihelion advance per period of Mercury's orbit*. It is conventional to express this as a figure for one century of Earth time. The period of Mercury's orbit is .241 Earth years, so in an Earth century, Mercury completes about 415 orbits. The total advance, in this case expressed in seconds of arc rather than radians, is then 43″. It is this quantity that was known in the nineteenth century to 1″ as an anomaly before the advent of general relativity, and which is predicted to this accuracy by general relativity.

We now turn to consider *null* geodesics. The geodesic equations are (18.1) and (18.2) as before together with

$$L = (1 - \frac{2M}{r})\dot{t}^2 - (1 - \frac{2M}{r})^{-1}\dot{r}^2 - r^2\dot{\phi}^2 = 0. \tag{18.14}$$

Solving for $\dot{r}$ as before, and introducing u, we find

$$(\frac{du}{d\phi})^2 = \frac{E^2}{J^2} - u^2 + 2Mu^3 \tag{18.15}$$

in place of (18.5). Again the 'Newtonian' equation is found by omitting the last term. The Newtonian solution is just

$$u = \frac{E}{J}\sin(\phi - \phi_0), \tag{18.16}$$

which is of course the equation of a straight line—in Newtonian theory light still travels on straight lines.

We choose $\phi_0 = 0$ so that the line is just $y = J/E$ and seek a solution to (18.15) of the form

$$u = \frac{E}{J}\sin\phi + V. \tag{18.17}$$

Substituting (18.17) into (18.15) and keeping only linear terms in V we find

$$\cos\phi\frac{dV}{d\phi} = -\sin\phi V + \frac{ME^2}{J^2}\sin^3\phi,$$

whence

$$V = \frac{3ME^2}{2J^2}(1 + \frac{1}{3}\cos 2\phi). \tag{18.18}$$

The effect of this is to produce a *deflection* from the straight line (18.16). To measure the deflection we need the limit of ϕ as $r \to \infty$, i.e. as $u \to 0$. If $\phi \to \phi_\infty \ll 1$ then from (18.17) and (18.18),

$$\frac{E}{J}\phi_\infty + \frac{3ME^2}{2J^2}\left(1 + \frac{1}{3}\right) = 0,$$

that is,

$$\phi_\infty = -\frac{2ME}{l}.$$

The total deflection will be *twice* this.

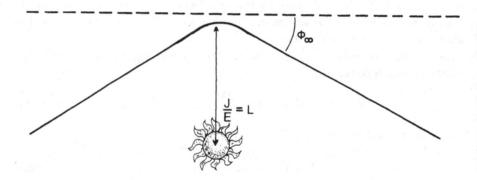

Figure 18.2. Deflection of light: the undeflected path is shown dashed.

So the total deflection, restoring the appropriate powers of G and c, is

$$\Delta\phi = \frac{4GM}{c^2 L},$$

where L is the closest approach of the light ray. We have already calculated $2GM/c^2$. To make $\Delta\phi$ as large as possible we consider a ray just grazing the sun, so that for L we must take the radius of the sun: $L = 6.96 \times 10^{10}$ *cm*. Expressed in minutes of arc, this gives a deflection of

$$\Delta\phi = 1.75'. \tag{18.19}$$

The most expedient way of measuring this quantity has been to use a solar eclipse to observe the apparent positions of stars near to the sun. It is also possible to use radio-astronomy, and observe the apparent position of radio sources occulted by the sun (with the advantage that you do not have to wait for an eclipse!). Either way there is agreement with (18.19) to about 1%. Extraordinary!

To lay the foundation for the next chapter, we must consider *radial* null geodesics, that is null geodesics with $J = 0$. The geodesic equations are

$$(1 - \frac{2M}{r})\dot{t} = E,$$

and

$$L = (1 - \frac{2M}{r})\dot{t}^2 - (1 - \frac{2M}{r})^{-1}\dot{r}^2 = 0,$$

so that

$$\dot{r}^2 = E^2, \tag{18.20}$$

i.e. $\dot{r} = \pm E$, and

$$\frac{dt}{dr} = \frac{\dot{t}}{\dot{r}} = \pm(1 - \frac{2M}{r})^{-1}. \tag{18.21}$$

From (18.20) for an *ingoing* null geodesic we have

$$r = r_0 - Es \tag{18.22}$$

and from (18.21),

$$t = -(r + 2M ln(r - 2M)) + constant. \tag{18.23}$$

For an outgoing null geodesic the signs are switched:

$$r = r_0 + Es \tag{18.24}$$

$$t = r + 2M ln(r - 2M) + constant \tag{18.25}$$

where in both expressions s is the affine parameter.

The question we have to consider is the following: a radially incoming geodesic will, according to (18.22), reach the Schwarzschild radius $r = 2M$ at a finite value of affine parameter s; but according to (18.23) as r approaches $2M$ the coordinate t goes to infinity. Thus the geodesic reaches $r = 2M$ at finite affine parameter but infinite t: so where does it go next? Similarly from (18.24) and (18.25) we see that a radially out-going null geodesic left $r = 2M$ a finite affine parameter distance in the past but at negative infinity in t, so where was it before that? These matters are addressed in the next chapter.

Exercises for chapter 18

[18.1] As a review of orbits in Newtonian gravity consider the Lagrangian

$$L = \frac{1}{2}(\dot{r}^2 + r^2\dot{\phi}^2) + \frac{M}{r}.$$

Show that $J = r^2\dot{\phi}$ and $h = \dot{r}^2 + r^2\dot{\phi}^2 - 2M/r$ are constants of the motion and deduce (18.8) and (18.9).

[18.2] Show that for a circular planetary orbit the angular velocity ω and the radius a are related by $\omega^2 a^3 = M$. (This is *Kepler's third law*, which is also valid in this form in Newtonian gravity).

[18.3] A radial light ray from the surface of a planet at radius b is intercepted by a space-craft in a circular orbit at radius a. Show that the observed red shift is given by

$$\frac{\nu_0}{\nu_E} = (1 - \frac{3M}{a})^{-\frac{1}{2}}(1 - \frac{2M}{b})^{\frac{1}{2}}.$$

[18.4] Show that there is a circular null geodesic at $r = 3M$ in the Schwarzschild solution.

[18.5] Show that, for a radial ingoing *space-like* geodesic γ in a surface of constant time t, r is given as a function of affine parameter s by

$$r^{\frac{1}{2}}(r - 2M)^{\frac{1}{2}} + 2M\ln(r^{\frac{1}{2}} + (r - 2M)^{\frac{1}{2}}) = s_0 - s.$$

Deduce that along γ the coordinate r decreases to $2M$ and then increases again.

[18.6] The object of this exercise is to find another test of general relativity based on a study of orbits—the *time delay*. Show that along a null geodesic, with the conventions of this chapter, we have

$$\frac{dt}{dr} = E(1 - \frac{2M}{r})^{-1}(E^2 - \frac{J^2}{r^2}(1 - \frac{2M}{r}))^{-\frac{1}{2}}$$
$$= \frac{E}{J}(1 - \frac{2M}{r})^{-1}(\frac{1}{b^2}(1 - \frac{2M}{b}) - \frac{1}{r^2}(1 - \frac{2M}{r}))^{-\frac{1}{2}},$$

where b is the radius of the closest approach. Show that the coordinate time from radius r to the point of closest approach is given approximately by

$$t = (r^2 - b^2)^{\frac{1}{2}} + M(2\log(\frac{r}{b} + (\frac{r^2}{b^2} - 1)^{\frac{1}{2}}) + (\frac{r - b}{r + b})^{\frac{1}{2}})$$

neglecting terms of order $(r_s/b)^2$. The first term is the answer in the absence of general relativity, so the second term, which represents a relativistic time delay, leads to a test of general relativity.

[18.7] (i) In Newtonian theory motion about an *oblate* star can be described by equation (18.8) with Mu replaced by the appropriate potential. Use the result of exercise [17.7] to show that for a nearly circular orbit of radius r there is a purely Newtonian precession (per orbit) proportional to r^{-2}. (ii) Combine (i) with exercise [18.2] to deduce that the advance per unit time is proportional to $r^{-7/2}$. (iii) A reactionary physicist claims from the above that Newton has explained away the mystery of Mercury and that Einstein's theory is redundant. Refute him thus: if this were the explanation for Mercury (radius R) then for *another* planet (at radius r) we should have from part (ii): advance per century $= (R/r)^{7/2} \cdot 43''$. But Venus is about twice as far away as Mercury, while its advance is roughly $9''$. What is Einstein's prediction? (The semi-major axis of Mercury's orbit is $57.9 \times 10^6 \, km$, whereas for Venus the figure is $108.2 \times 10^6 \, km$.)

[18.8] Show that the equation for null geodesics in the equatorial plane of the Schwarzschild solution can be written

$$(\frac{du}{d\phi})^2 = \frac{E^2}{J^2} - u^2 + 2u^3.$$

Deduce that for a suitable value of E/J there are null geodesics on which

$$\frac{1 - 3u}{(\sqrt{3} \pm \sqrt{1 + bu})^2} = Ae^\phi$$

for arbitrary constant A. Describe their behaviour as $\phi \to -\infty$ for $A = 0$ or $A > 0$. (Compare with exercise 18.4.)

19   The extended Schwarzschild solution

THE DEFINITION of a manifold was sketched out in chapter 4 in terms of a system of overlapping coordinate patches. But in deriving the Schwarzschild solution we set to one side the question of the underlying manifold and concentrated on one primary coordinate patch chosen to exhibit the assumed symmetries.

We tacitly assumed that this patch was one of a number covering the underlying manifold; but we made no assumptions about the manifold since we had no intuition to 'guide' us, no notion as to what would constitute sensible assumptions.

Now the geodesics on a manifold (with metric) are intrinsic: they do not particularly care about the choice of coordinates. Thus once we have the metric in one patch, we can find the geodesics in that patch and use them as a guide to lead us to other patches. This is precisely what happens here. We derived the Schwarzschild metric as

$$ds^2 = (1 - \frac{2M}{r})dt^2 - (1 - \frac{2M}{r})^{-1}dr^2 - r^2(d\theta^2 + \sin^2\theta d\theta^2) \tag{19.1}$$

on a patch U_1 specified by

$$-\infty < t < \infty, \quad 2M < r < \infty, \quad 0 \leq \theta \leq \pi, \quad 0 \leq \phi < 2\pi. \tag{19.2}$$

(Strictly, this patch does not cover the north and south poles, but this minor detail is unimportant for the discussion). We impose the restriction $r > 2M$, since certainly $r = 2M$ is not allowed because the metric is singular there, and then $r < 2M$ is a disconnected region.

We expect to find new patches to cover the parts of the space corresponding to $r < 2M$. From the concluding remarks of chapter 18 we know that radially incoming null geodesics run off U_1 at infinite t at a finite affine parameter into the future, and that radially outgoing null geodesics run off U_1 at negative infinite t at a finite affine parameter into the past.

Also from exercise [18.5] we know that radial space-like geodesics run off U_1 with any value of t. We therefore want new patches to the future and past (and possibly 'sideways' as well) for these geodesics to run onto.

To find the new coordinates, we return to the discussion at the end of chapter 18. There we saw that a radially incoming null geodesic has

$$t + r + 2M\ell n(r - 2M) = \text{constant}.$$

Therefore, if we introduce a new coordinate v by

$$v = t + r + 2M\ell n(r - 2M), \tag{19.3}$$

then the radially incoming null geodesic is given by

$$v = \text{constant}, \quad r = r_0 - Es.$$

The problem has now been removed. The coordinate t became infinite at $r = 2M$, so we eliminate t in favour of v which is *finite* on the incoming geodesics. In the coordinates (v, r, θ, ϕ) the metric (19.1) is

$$ds^2 = (1 - \frac{2M}{r})dv^2 - 2dvdr - r^2(d\theta^2 + \sin^2\theta d\phi^2), \tag{19.4}$$

and the allowed range of coordinates is now

$$-\infty < v < \infty, \quad 0 < r < \infty, \quad 0 < \theta < \pi, \quad 0 \leq \phi < 2\pi. \tag{19.5}$$

These coordinates and the analogous ones of (19.7) are referred to as the *Eddington-Finkelstein* coordinates. Calling this patch U_2, we see that it in fact includes all of U_1, but extends it across the surface at $r = 2M$ towards the future. Incoming null geodesics now cross $r = 2M$ at finite v and continue on down to $r = 0$ arriving there at a finite affine parameter. Also the metric itself is now regular at $r = 2M$, although still singular at $r = 0$.

The relation between U_1 and U_2 is sketched in figures (19.1a,b,c). Each point of the figures represents a sphere whose radius is the corresponding value of r.

In figure 19.1a we show the incoming and outgoing null geodesics in (t, r) coordinates. For large r they are approximately straight but as they approach $r = 2M$ they veer up to infinite t or down to negative infinite t.

The incoming geodesics have constant v, so if we straighten them out we get the picture in figure 19.1b. New points have been added beyond $r = 2M$, effectively beyond '$t = +\infty$'.

In this picture, the outgoing null geodesics still run off the patch, now at $v \to -\infty$. We can cope with these by using (18.25) instead of (18.23). That is, we introduce a new coordinate u by

$$u = t - r - 2M\ln(r - 2M), \tag{19.6}$$

and eliminate t in favour of u:

$$ds^2 = (1 - \frac{2M}{r})du^2 + 2dudr - r^2(d\theta^2 + \sin^2\theta d\phi^2). \tag{19.7}$$

This gives a patch U_3 with

$$-\infty < u < \infty, \quad 0 < r < \infty,$$

and with θ, ϕ as before.

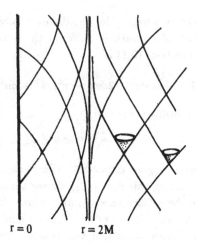

Figure 19.1a. Incoming and outgoing geodesics in (t, r) coordinates.

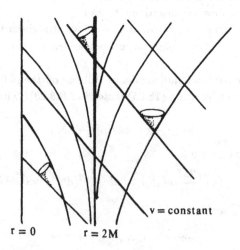

Figure 19.1b. Incoming and outgoing geodesics in (v, r) coordinates.

Now the outgoing null geodesics have a 'home'. They come from $r < 2M$ at a constant u. Again U_3 includes all of U_1 but extends it now across $r = 2M$ into the past. Also, the metric is again regular at $r = 2M$ but singular at $r = 0$.

The process of extension is not quite finished as we still have the radial space-like geodesics with constant t. We recall from exercise [18.5] that these are determined by

$$\dot{t} = 0, \quad (1 - \frac{2M}{r})^{-1}\dot{r}^2 = 1.$$

On such a geodesic r decreases to $2M$ and then increases again. Such a geodesic runs off the bottom of figure (19.1b) and runs off the top of figure (19.1c)! To see where it goes we put these two figures together, with all null geodesics at $45°$ as shown in figure 19.2.

Now the region in the top is the part of U_2 not in U_1 and the region in the bottom is the part of U_3 not in U_1. Both of these correspond to the metric (19.1) but in a region with $0 < r < 2M$. Note that this effectively changes the signature of (19.1) from $(+ - - -)$ to $(- + - -)$; that is, t becomes a space-like coordinate while r becomes time-like. These two regions can now be extended to the left, since they must clearly be symmetric into a region U_4 which is another copy of U_1. There is then a cross-over point in the middle which has not so far appeared in any of the coordinate patches. We must check that this cross-over point can be covered by some patch. Suppose this done, then the geodesics at constant t come in from decreasing r in U_1, pass through $r = 2M$ at the cross-over and go into U_4 with r increasing again.

This time the extension is complete. All geodesics stay on one patch or another to infinite values of their affine parameter except for the ones which reach $r = 0$. We recall that each form of the metric is singular at $r = 0$, and we would be inclined to regard the space-time as singular there. However we know by experience that a singularity in one of the metric components may well be removed by a suitable coordinate transformation. To say with confidence that the space-time is singular we need some appropriate quantity to be infinite in *all* coordinate systems. That is, we seek a scalar quantity that becomes infinite. We cannot use the Ricci scalar, or any quantity constructed from the Ricci tensor since these are zero by assumption. The simplest thing to try is $R_{abcd}R^{abcd}$. Calculation reveals that this is given by

$$R_{abcd}R^{abcd} = \frac{48M^2}{r^6}$$

which is indeed singular at $r = 0$ in all coordinate systems. Thus the Schwarzschild solution has a singularity at $r = 0$, which we know from our investigation actually happens in 'two places'.

This analysis is very important: it forms the basis for the discussion of black holes and the process of gravitational collapse.

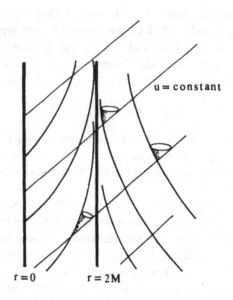

Figure 19.1c. Incoming and outgoing geodesics in (u, r) coordinates.

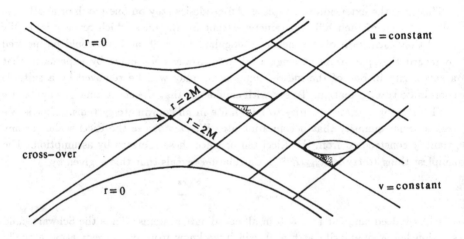

Figure 19.2. Fitting together U_2 and U_3. The space on the left is filled by U_4.

Exercises for chapter 19

[19.1] A null surface is a surface Σ whose normal, say n_a, is null. Suppose in a co-ordinate system (x^0, x^i) $i = 1, 2, 3$ the surfaces of constant x^0 are null. Show that the component g^{00} of the metric is zero, that the components g^{0i} are not all zero and that the matrix of components g_{ij} is degenerate. Deduce that in the Schwarzschild solution the surfaces of constant v or of constant u are null surfaces, as is the surface $r = 2M$.

[19.2] Show that a null curve in a null surface is necessarily a null geodesic.

[19.3] If a solution ϕ of the wave equation

$$\Box \phi \equiv g^{ab} \nabla_a \nabla_b \phi = 0$$

is discontinuous across a surface Σ, show that Σ must be a null surface. Show that the same result holds for Maxwell's vacuum equations.

[19.4] Show that the coordinate transformation

$$r = \frac{(2R + M)^2}{4R}$$

puts the metric of the constant t-surface in the Schwarzschild solution into the form

$$ds^2 = (1 + \frac{M}{2R})^4 (dR^2 + R^2(d\theta^2 + \sin^2\theta d\phi^2)).$$

Now show that inversion in the sphere $R = M/2$, i.e. the transformation

$$R \to \hat{R} = \frac{M^2}{4R},$$

is an isometry of this metric. Sketch out a schematic picture of the $t = $ constant surface paying careful attention to its asymptotes, and use the result to explain the behaviour of the space-like geodesics in exercise [18.5].

[19.5] In chapter 17 we saw that $E = (1 - \frac{2M}{r})\dot{t}$ is constant along null geodesics. Where in figures 19.1 are the null geodesics with $E = 0$?

[19.6] Can you find suitable coordinates to include the cross-over point in figure 19.2? The main idea is to introduce u and v and eliminate r.

[19.7] Two metrics $(g, \hat{g})$ are said to be 'conformal' if there is a function Φ for which $\hat{g}_{ab} = \Phi^2 g_{ab}$. The transition $g \to \hat{g}$ is locally no more than an expansion, and angles are thus preserved. Of deeper significance is the fact (cf. exercise 9.3) that the null geodesics (light cones) are conformally invariant. The 'causal structure' (which events can physically influence which other events) is thus also preserved. We can arrange for Φ to compress distance increasingly the further away we go, and render infinity 'finite'. As a simple example take Minkowski space-time in spherical polar coordinates:

$$ds^2 = dt^2 - dr^2 - r^2(d\theta^2 + \sin^2\theta d\phi^2).$$

First introduce null coordinates $v = t + r$, $u = t - r$ and re-express the metric in terms of these, the whole space-time being given by $-\infty < v < \infty$, $-\infty < u < \infty$, $v \geq u$. Next introduce p, q defined by $v = \tan p$, $u = \tan q$ and deduce that the original space-time is conformal to

$$d\hat{s}^2 = 4dpdq - sin^2(p - q)(d\theta^2 + \sin^2\theta d\phi^2)$$

with

$$-\frac{\pi}{2} < p < \frac{\pi}{2}, \quad -\frac{\pi}{2} < q < \frac{\pi}{2}, \quad p > q.$$

[19.8] Write down the null geodesic equations for the two-dimensional indefinite metric

$$ds^2 = z^2 dt^2 - dz^2 \quad 0 \leq z < \infty, \quad -\infty < t < \infty.$$

Show that ze^t is constant along incoming null geodesics (i.e. null geodesics with $dz/dt < 0$). Show that ze^{-t} is constant along outgoing null geodesics. For a null geodesic γ emanating from a point with $z = z_0$, and with initial velocity $\dot{z}_0 = [dz/ds]|_{s=0}$ where s is an affine paramter, show that

$$z = (z_0^2 + 2z_0\dot{z}_0 s)^{1/2}.$$

Show that γ reaches $z = 0$ at a finite affine paramter value; what is the value of t for this value of s? Consider the coordinate transformation given by

$$T - X = -ze^t, \quad T + X = ze^{-t}.$$

What is ds^2 expressed in (T, X) coordinates? What part of the (T, X) plane is represented by the (t, z) half-plane?

20 Black holes and gravitational collapse

Quicquid excessit modum pendet instabili loco. (*Whatever has exceeded its proper bounds is in a state of instability.*)

—Seneca

GRAVITATIONAL collapse, leading to the formation of black holes and ultimately to space-time singularities, constitutes one of the most interesting   and important consequences of general relativity. We consider a spherically symmetric star of radius say r_0. The gravitational field of the star is represented by the Schwarzschild solution at $r > r_0$ and by some spherically symmetric solution with matter at $r < r_0$ (see chapter 21 for an example of an 'interior' solution).

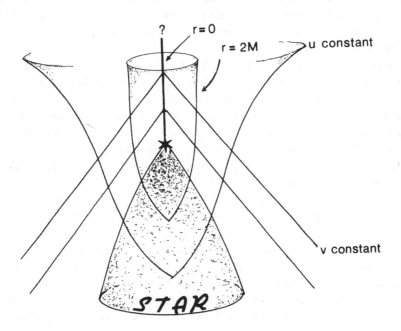

Figure 20.1. The collapse of a spherically symmetric star.

During its normal life-time the star generates energy by nuclear fusion in its interior. The tendency of the star to collapse under gravity is opposed by the pressure of the matter and energy emerging from the interior. When the nuclear fuel is spent the star contracts. If the star is sufficiently massive, the gravitational forces overpower the pressure, and the star collapses down to a point of infinite density at $r = 0$.

Outside the star the metric is given by the Schwarzschild solution. Note that the world lines of points on the surface of the star stay inside the local light cones. The null surface at $r = 2M$ appears outside the star as the history of a flash of light emitted from the centre of the star. If we imagine a point on the surface of the star emitting equally spaced flashes of light then these are seen by a distant observer as more and more separated in time. The flash emitted as the surface crosses $r = 2M$ is then infinitely delayed. The fate of the star after it crosses $r = 2M$ is completely inaccessible to an observer who remains outside. For this reason the surface at $r = 2M$ is known as an *absolute event horizon*.

While an external observer never sees the star cross the event horizon, it should not be thought that he will always see the star as hovering just outside. This is because the gravitational red shift will make light from the star exponentially red-shifted, i.e. the star will become dark exponentially fast as it approaches the event horizon. Put another way, light emitted from the surface of the star in the last second of proper time as it crosses the absolute event horizon must last the whole infinite future of the external region and so is spread very thin! Nonetheless, since the external metric is the Schwarzschild solution, planets can continue to orbit the departed star so that its gravitational influence is still felt.

In this state, the star has become a black hole: 'black' for reasons given above, and a 'hole' because it is possible for other objects now to cross the horizon but never to return. As we shall see in exercise [20.1], any time-like world line once it has crossed the horizon will inevitably encounter the singularity $r = 0$. Any observer following such a world line will be destroyed by the large tidal forces near the singularity.

The area of the event horizon, i.e. the area of any space-like two-surface lying in $r = 2M$ and going round once, is $4\pi(2M)^2 = 16\pi M^2$. Thus the area gives a measure of the *mass* of the black hole.

An artificial feature of the discussion so far is the assumption of *spherical symmetry*. It is an important question how far the main features of this picture persist for an *asymmetric* star.

A detailed theory of the structure and evolution of stars has established that large classes of stars may be expected to collapse gravitationally at the end of their evolution. Indeed, there exist theorems to the effect that, under fairly general circumstances, when the collapse has gone past a certain point the formation of a singularity is *inevitable*. Finally, a stability analysis of the Schwarzschild solution shows that small perturbations will settle down.

One can say with some confidence that many stars will collapse and, if general relativity is valid in the conditions that follow, many singularities will form. Modifications of general relativity may enable the singularities to be avoided or mollified, but such considerations are inevitably speculative. However, if black holes always form then one is able in a sense to postpone worrying about singularities. If every singularity formed in collapse is 'decently clothed' by an event horizon, then no singularity will be visible to an outside observer. The idea that this is what happens is known as the 'cosmic censorship hypothesis'.

There is an accumulation of evidence for the cosmic censorship hypothesis including the stability of the Schwarzschild solution mentioned above. A proof of it under fairly general conditions may be far off however, and it has some claim to being for the moment one of the most important unsolved problems in relativity.

Putting to one side the question of whether black holes always form in collapse, one can study them in their own right. The great astrophysicist S. Chandrasekhar has called this the 'purest branch of physics' since the objects of study are 'made' of space and time, and the details of the behaviour of matter, so crucial for example in the theories of stellar structure and evolution, are not so important—and indeed are nearly irrelevant. One result from this study, which we single out for mention, is that the *general* isolated stationary black-hole in empty space is described by the *Kerr* solution. As we shall see in chapter 22 this has two parameters identifiable as *mass* and *angular momentum*. Thus the general black hole, assuming it settles down eventually, is characterized by just these two quantities.

Exercises for chapter 20

[20.1] Show that any observer following an incoming time-like path must reach $r = 0$ within a proper time $2\pi M$ after passing $r = 2M$.

[20.2] From the metric written in the usual (t, r, θ, ϕ) coordinates we can easily read off the Schwarzschild mass M. How can we do this in a more geometric (coordinate independent) way? At a point p we may construct the sphere of symmetry S_p through it by demanding that the geometry is uniform on S_p. Let $A(p) =$ area of S_p. Find a formula for M in terms of this area function and its radial rate of change with respect to proper distance. In fact this formula is even valid *inside* a spherical star. M is, of course, no longer constant but rather represents the mass interior to the sphere at which the formula is evaluated.

[20.3] The unfortunate physicist of exercise [18.7 iii] refuses to mend his ways and now claims that the black holes are the delusions of deranged minds. He is put on trial by a group of relativists, and sentenced to death by radial injection into a black hole. Upon nearing the event horizon he decides to repent by radio to the zealots, who are at rest far from the black hole. They find that his message is red-shifted in proportion to $exp(Kt)$, where t is their proper time. Show that they can deduce the mass of the executioner.

[20.4] In exercise [19.7] we saw how a conformal factor could be used to compress Minkowski space-time and render infinity 'finite'. The same can be done for the Schwarzschild geometry. First express it in terms of the double null coordinates u and v, as defined in chapter 19 (cf. exercise 19.6). To eliminate the remaining singular factor $(1 - \frac{2M}{r})$ introduce new coordinates

$$V = exp(\frac{v}{4M}), \quad U = -exp(\frac{-u}{4M})$$

Lastly, to compress the space-time, introduce new coordinates p,q by analogy with exercise [19.7].

21   Interior solutions

21.1   Potential theory revisted

A CLASSICAL PROBLEM in Newtonian gravitational theory is the calculation of the gravitational potential of a *uniform sphere of mass*. Thus if the sphere is of radius R and constant density ρ_o we require in Newton's theory a spherically symmetric solution of the Newton-Poisson equation

$$\nabla^2 \Phi = 4\pi G\rho \qquad (21.1.1)$$

subject to

$$\rho = 0 \quad (r > R) \qquad (21.1.2)$$

$$\rho = \rho_o \quad (r \leq R) \qquad (21.1.3)$$

such that $\Phi(r) \to 0$ for large r. The solution to this system of relations is well known to be given by

$$\Phi = -\frac{GM}{r} \quad (r > R) \qquad (21.1.4)$$

$$\Phi = \frac{GM}{2R^3}(r^2 - 3R^2) \quad (r \leq R) \qquad (21.1.5)$$

where

$$M = \frac{4}{3}\pi R^3 \rho_o \qquad (21.1.6)$$

is the total mass of the sphere. It is a straightforward exercise to verify that at the boundary $r = R$ the 'interior' and 'exterior' potentials agree with one another, as do their first derivatives.

Thus summing up, we see that (21.1.4) and (21.1.5) together represent the complete gravitational field, in Newton's theory, for a uniform sphere of mass. Is there any such analogy to this 'complete' field in Einstein's theory? If so, it would evidently be of considerable importance since it would form the starting point for a theory of relativistic stellar structure. In fact, such a solution *does* exist, and here we shall describe it.

21.2   Self-gravitating fluids in Newtonian theory

In Einstein's theory, before proceeding to solve the field equations, we must take a view as to the constitution of the *source* of the gravitational field. It is instructive therefore to develop this point first in the context of the Newtonian theory, and specifically to consider the case of *self-gravitating fluid in hydrostatic equilibrium*. Many of the ideas arising here carry over to the relativistic theory.

Now in Newton's theory the equations governing a self-gravitating fluid are

$$\rho[\dot{V}_i + V_j \nabla_j V_i] = -\nabla_i p - \rho \nabla_i \Phi, \tag{21.2.1}$$

together with the Newton-Poisson equation, as discussed in §2.5. But for a fluid in static equilibrium the velocity three-vector V_i vanishes, so we are left with

$$\nabla_i p = -\rho \nabla_i \Phi, \tag{21.2.2}$$

representing the balance of pressure and gravitation. So in Newton's theory a self-gravitating fluid in static equilibrium satisfies (21.1.1) and (21.2.2).

In the spherically symmetric case for any function $f(r)$ we have

$$\nabla_i f(r) = \frac{x_i}{r} f' \tag{21.2.3}$$

where $r^2 = x_i x_i$ and the dash denotes d/dr, and also

$$\nabla_i \nabla_i f(r) = f'' + \frac{2}{r} f'. \tag{21.2.4}$$

Thus if p, ρ, and Φ are functions of r alone then for (21.1.1) and (21.2.2) we have

$$\Phi'' + \frac{2}{r} \Phi' = 4\pi G \rho \tag{21.2.5}$$

and

$$p' = -\rho \Phi'. \tag{21.2.6}$$

The object now is to eliminate Φ so as to obtain an equation relating ρ and p. To be definite let us suppose furthermore that we want fields such that ρ, p, and Φ and their derivatives are non-singular throughout the 'star' and its environment. If we multiply (21.2.5) by r^2 and integrate, it follows, after some elementary manipulation, that

$$\Phi' = \frac{4\pi G}{r^2} \int_0^r \rho(\xi) \xi^2 \, d\xi + \frac{k}{r^2} \tag{21.2.7}$$

where k is a constant; but k vanishes by our assumption that $\Phi'(0)$ is non-singular. Thus if we write

$$m(r) = 4\pi \int_0^r \rho(\xi) \xi^2 \, d\xi \tag{21.2.8}$$

for the total mass up to the radius r, we have

$$\Phi' = \frac{G m(r)}{r^2}. \tag{21.2.9}$$

Finally therefore by substitution of (21.2.9) in (21.2.6) we get

$$p' = \frac{-G \rho m(r)}{r^2}. \tag{21.2.10}$$

This is the fundamental equation of hydrostatic equilibrium for a spherically symmetric self-gravitating system of fluid in Newton's theory, and is the starting point for the study of non-relativistic stellar structure.

21.3   The Oppenheimer-Volkoff equation

Now suppose we have a sperically symmetric star of radius R and mass M in general relativity. Then for $r > R$ the gravitational field is given by the Schwarzschild solution

$$ds^2 = (1 - \frac{2GM}{r})dt^2 - (1 - \frac{2GM}{r})^{-1}dr^2 - r^2 d\Omega^2, \qquad (21.3.1)$$

whereas for the interior $r \leq R$ we expect a solution of the form

$$ds^2 = e^{2\Phi}dt^2 - e^{2\Psi}dr^2 - r^2 d\Omega^2 \qquad (21.3.2)$$

where Φ and Ψ are functions of r alone, with a stress-energy tensor of the form

$$T_{ab} = (\rho + p)u_a u_b - p g_{ab} \qquad (21.3.3)$$

where ρ and p are functions of r alone. The fluid velocity four-vector u_a must be given by

$$u_a = \xi_a/(\xi_b \xi^b)^{1/2} \qquad (21.3.4)$$

where $\xi^a = (1, 0, 0, 0)$ is the time-like Killing vector.

Given the metric (21.3.2) it is a straightforward matter to work out Einstein's equations for the stress-energy tensor (21.3.3). Assuming, as in the Newtonian case, that the various quantities of relevance are not singular at the origin, we can integrate Einstein's equations to find for $\Psi(r)$ that

$$e^{2\Psi} = [1 - \frac{2Gm(r)}{r}]^{-1} \qquad (21.3.5)$$

where the function $m(r)$ is given, by definition, as in the Newtonian case by (21.2.8). For $\Phi(r)$ instead of (21.2.9) we have

$$\Phi' = \frac{Gm(r)}{r^2}[1 + \frac{4\pi r^3 p}{m(r)}][1 - \frac{2Gm(r)}{r}]^{-1}. \qquad (21.3.6)$$

And finally instead of (21.2.10) we have

$$p' = \frac{-G\rho m(r)}{r^2}[1 + \frac{p}{\rho}][1 + \frac{4\pi r^3 p}{m(r)}][1 - \frac{2Gm(r)}{r}]^{-1}. \qquad (21.3.7)$$

This is the equation of hydrostatic equilibrium for a spherically symmetric system in general relativity, and is known as the *Oppenheimer-Volkoff equation*. The basic form of the Newtonian equation (21.2.10) is maintained, but some 'correction' terms appear.

Like its Newtonian analogue, the Oppenheimer-Volkoff equation is a first order non-linear equation for $p(r)$. Once we are given $\rho(r)$ and a suitable boundary condition, e.g. $p(R) = 0$, then (21.3.7) can be solved for $p(r)$, and (21.3.6) can be solved for $\Phi(r)$. This is the approach we shall use in the next section in the case of constant density. In general however $\rho(r)$ is not given, but rather p as a function of ρ, e.g. a relation such as $p = k\rho^n$. Such a relation will be given in advance through physical considerations, then (21.3.7) solved for $p(r)$ subject to an appropriate boundary condition. Exact solutions to (21.3.7) are very difficult to obtain; but numerical solutions can readily be computed and these certainly suffice for practical purposes.

21.4   Schwarzschild's interior solution

In the special and very idealized case of a star of *uniform density*, equations (21.3.6) and (21.3.7) can be integrated exactly so as to yield very neat expressions for the metric components and the pressure.

For a spherically symmetric star of uniform density ρ is given by (21.1.2) and (21.1.3). And thus, according to (21.2.8), $m(r)$ is determined to be

$$m(r) = M \quad (r > R) \tag{21.4.1}$$

$$m(r) = Mr^3/R^3 \quad (r \leq R) \tag{21.4.2}$$

where R is the radius of the star, and $M = \frac{4}{3}\pi R^3 \rho_0$ is the total mass.

Equation (21.3.7) can be integrated, with the boundary condition $p(R) = 0$, to yield the following expression

$$p(r) = \rho_0 [\frac{(1 - 2GM/R)^{1/2} - (1 - 2GMr^2/R^3)^{1/2}}{(1 - 2GMr^2/R^3)^{1/2} - 3(1 - 2GM/R)^{1/2}}] \tag{21.4.3}$$

for $r < R$, with $p = 0$ for $r > R$. And for $\Phi(r)$ we obtain

$$e^{2\Phi} = [\frac{3}{2}(1 - 2GM/R)^{1/2} - \frac{1}{2}(1 - 2GMr^2/R^3)^{1/2}]^2 \tag{21.4.4}$$

for $r < R$, and

$$e^{2\Phi} = 1 - 2GM/r \text{ for } r > R.$$

These expressions may appear a trifle complicated at first glance, but on closer inspection they will be seen to be very natural expressions that reduce readily to the corresponding Newtonian formulae in the appropriate limit.

Exercises for chapter 21

[21.1] For a spherical relativistic star of constant density show that $GM < \frac{4}{9}R$.

[21.2] By use of the Ricci tensor expressions already given for a general spherically symmetric metric, calculate Einstein's equations for the metric (21.3.2) with the stress tensor given by (21.3.3) and (21.3.4). Show that (21.3.5) and (21.3.6) follow directly from the field equations; whereas (21.3.7) is equivalent to the contracted Bianchi identity.

[21.3] Verify that (21.3.7) can be integrated, in the case of constant density, to yield (21.4.3).

[21.4] Construct a solution of Einstein's equations corresponding to a spherical relativistic star with two 'layers', an inner layer with constant density ρ_o, and an outer layer with constant density ρ_1. Does this star also satisfy the inequality $GM < \frac{4}{9}R$?

[21.5] Show that the central pressure of a constant density spherical relativistic star is greater that that of a Newtonian star of the same 'build'. For a star with the mass and radius of the sun, but idealized to have uniform density, calculate and compare the central pressures in the Newtonian and general relativistic theories.

[21.6] Verify that (21.4.3) and (21.4.4) reduce in the limit to the appropriate Newtonian formulae. In the Newtonian case show that the central pressure P_c satisfies the inequality

$$P_c > \frac{1}{8\pi}\frac{GM^2}{R^4},$$

regardless of the relation between p and ρ, apart from the general conditions of equilibrium. Do you expect a similar relation to hold in Einstein's theory?

[21.7] Calculate the Weyl tensor associated with the constant density interior Schwarzschild solution. What about the more general case of a system satisfying the Oppenheimer-Volkoff relations?

22   The Kerr solution

'If only it weren't so damnably difficult to find exact solutions!'
—Albert Einstein (undated letter to M. Born, c. 1936)

NO SINGLE theoretical development in the last three decades has had more influence on gravitational theory than the discovery of the *Kerr solution* in 1963. The Kerr metric is a solution of the vacuum field equations. It is a generalization of the Schwarzschild solution, and represents the gravitational field of a special configuration of rotating mass, much as the external Schwarzschild solution represents the gravitational field of a spherical distribution of matter.

However, unlike the Schwarzschild case, no simple non-singular fluid 'interior' solution is known to match onto the Kerr solution. There is, nevertheless, no reason a *priori* why such a solution shouldn't exist.

Fortunately such speculations are in some respects beside the point, since the real interest in the Kerr solution for many purposes is its characterization of the final state of a *black hole*, after the hole has had the opportunity to 'settle down' and shed away (via gravitational radiation and other processes) eccentricities arising from the structure of the original body that formed the black hole.

To put the matter another way, suppose someone succeeded in exhibiting a good fluid interior for the Kerr metric. Well, that would be in principle very interesting; but there is no reason to believe that naturally occurring bodies (e.g. stars, galaxies, etc.) would tend to fall in line with that particular configuration. On the other hand, following the gravitational collapse of any rotating configuration of mass, the resulting black hole (as it settles down to an equilibrium state) moves into the Kerr configuration as regards its external field. This scenario cannot necessarily be taken as the gospel, but there are numerous theoretical arguments of varying degrees of rigour, generality, and sophistication that support this point of view, all therefore confirming the special significance of the Kerr solution as an object of study.

22.1   Boyer-Lindquist coordinates

The Kerr solution depends on two parameters: the total mass M and the total angular momentum J. For practical purposes it is convenient to introduce a *specific angular momentum* parameter $a = J/M$. As with the Schwarzschild solution, the Kerr metric can be put into several varying forms, depending on the choice of coordinates being used locally to cover a particular part of the manifold.

First we consider the so called 'Boyer-Lindquist' form of the metric. At first glance, it is exceedingly complicated compared with anything we have seen so far:

$$ds^2 = dt^2 - 2Mr\Sigma^{-1}(dt - a\sin^2\theta d\phi)^2 - \Sigma(\Delta^{-1}dr^2 + d\theta^2) - (r^2 + a^2)\sin^2\theta d\phi^2 \quad (22.1.1)$$

where

$$\Sigma = r^2 + a^2\cos^2\theta \quad (22.1.2)$$

and

$$\Delta = r^2 - 2Mr + a^2. \quad (22.1.3)$$

Do not, however, be put off by first appearance: the metric becomes more surveyable with familiarity, and pays back the time invested in its study with good interest. Rearranging the terms in (22.1.1) slightly we see equivalently that it can be written in the form

$$ds^2 = \Sigma^{-1}\Delta(dt - a\sin^2\theta d\phi)^2 - \Sigma(\Delta^{-1}dr^2 + d\theta^2) - \Sigma^{-1}\sin^2\theta[(r^2 + a^2)d\phi - adt]^2. \quad (22.1.4)$$

In either case we see that as $a \to 0$ we recover the Schwarzschild metric

$$ds^2 = (1 - 2M/r)dt^2 - (1 - 2M/r)^{-1}dr^2 - r^2(d\theta^2 + \sin^2\theta d\phi^2). \quad (22.1.5)$$

The metric (22.1.4) is evidently singular at $\Sigma = 0$, corresponding to the 'true' singularity at $r = 0$ in the Schwarzschild solution. Now on reflection it may seem slightly odd that the solution is singular for $r = 0$ and $\theta = \frac{1}{2}\pi$, but not for $r = 0$ with other values of θ. But this apparent paradox arises from a naive conception of the meaning of the coordinate r. The surfaces of constant r are in fact a family of *confocal oblate ellipsoids*, and the ellipsoid at $r = 0$ degenerates into a *disk*, the boundary of this disk being given by $\theta = \frac{1}{2}\pi$. Thus the Kerr solution has a 'ring' singularity, rather than a 'point' singularity. These aspects of the solution can be seen best when the metric is expressed in quasi-Minkowskian or 'Kerr-Schild' coordinates, as we shall see shortly.

Apart from the true 'ring' singularity at $\Sigma = 0$ the metric (22.1.4) also possesses a pair of spurious pseudo-singularities (analogous to the surface $r = 2M$ for Schwarzschild) at the real roots of $\Delta = 0$. These exist only when $a^2 \leq M^2$, and are given by $r = r_\pm$ where

$$r_\pm = M \pm (M^2 - a^2)^{1/2}.$$

In fact, the manifold is perfectly regular at $r = r_\pm$, as can be seen by changing to a new set of coordinates (Kerr-Eddington coordinates).

Thus the Boyer-Lindquist system, though very natural for the study of asymptotic properties of the Kerr solution, needs to be supplemented by further coordinatizations in order that the manifold should be properly built up and understood.

22.2   Kerr-Eddington coordinates

To eliminate the pseudo-singularities shown in the Boyer-Lindquist system we introduce a new set of coordinates (T, R, Θ, Φ) related to the Boyer-Lindquist set (t, r, θ, ϕ) by the following transformation:

$$dT = dt + 2Mr\,dr/\Delta$$
$$dR = dr$$
$$d\Theta = d\theta$$
$$d\Phi = d\phi - a\,dr/\Delta \qquad\qquad (22.2.1)$$

where $\Delta(r) = r^2 - 2Mr + a^2$. There is nothing particularly special about the fact that the transformation is written here in terms of differential forms: this merely makes rewriting the expression for ds^2 easier, and also avoids some slightly awkward (though perfectly manageable) integration. Thus we imagine the Kerr manifold as containing at least two patches, covered respectively by the (T, R, Θ, Φ) system and the (t, r, θ, ϕ) system, and in the overlap of these two patches the coordinates are identified according to the scheme (22.2.1). In fact, we have been given the Kerr manifold initially only as far as one patch (viz., the Boyer-Lindquist patch) and we are using the transformation (22.2.1) to 'extend' the manifold, i.e. to 'discover' more of it.

The result is the form of the metric originally discovered by Kerr, which for clarity here we shall call the Kerr-Eddington form, by virtue of its comparison with the Schwarzschild metric in the Eddington-Finkelstein form, to which it reduces in the limit as $a \to 0$. The Kerr-Eddington form of the metric is :

$$ds^2 = dT^2 - dR^2 - 2a\sin^2\Theta\,dR\,d\Phi - \Sigma\,d\Theta^2 - (R^2 + a^2)\sin^2\Theta\,d\phi^2$$
$$- (2MR/\Sigma)(dR + a\sin^2\Theta\,d\Phi + dT)^2. \qquad\qquad (22.2.3)$$

where, as before, Σ is defined by

$$\Sigma(R, \Theta) = R^2 + a^2\cos^2\Theta. \qquad\qquad (22.2.4)$$

The key feature to observe here is that the metric is *manifestly non-singular* at $\Delta = 0$, but retains the singularity at $\Sigma = 0$.

22.3   The time and angular Killing vectors

Note that in both the Boyer-Lindquist as well as the Kerr-Eddington forms of the metric the time and angular Killing vectors can be written in the form

$$T^a = (1, 0, 0, 0), \qquad\qquad (22.3.1)$$
$$\Phi^a = (0, 0, 0, 1).$$

These Killing vectors generate the stationary and axial symmetries of the solution.

It is useful, for future reference, to compute the norms of these Killing vectors. Writing the results in the r,θ system (we could equally well use the R,Θ system, simply substituting R,Θ for r,θ) we obtain:

$$T^a T_a = 1 - \frac{2M}{r}\left(\frac{r^2}{r^2 + a^2 \cos^2\theta}\right), \qquad (22.3.2)$$

and

$$\Phi^a \Phi_a = -r^2 \sin^2\theta - a^2\left[\sin^2\theta + \frac{2Mr\sin^4\theta}{r^2 + a^2\cos^2\theta}\right]. \qquad (22.3.3)$$

An interesting point immediately arises: neither of these expressions necessarily has a definite sign. Certainly $T^a T_a > 0$ for large r, as we would expect. But for sufficiently small r it goes negative! And what about $\Phi^a \Phi_a$? *Prima facie* it looks negative definite (as we would expect, in our signature, for an 'axial' Killing vector). However, there is a tacit assumption here, which need not be made: the coordinate r doesn't have to be positive! Certainly, as r is large it corresponds roughly to a spherical radial coordinate; but $r = 0$ corresponds not to a point but to an entire disk of radius a. If we approach the disk from above (say) and go through it, then we reach a region of negative r. In effect we are extending the manifold into another 'sheet', where r is negative. And in this negative r part of the solution $\Phi^a \Phi_a$ can be positive (as we see from the expression above) provided $|r|$ is not too large. Clearly the behaviour of the Kerr solution is unusual in the neighbourhood of its singularity. Now since Φ^a is associated with an angular, hence *periodic* symmetry, it follows that its orbits are closed; and thus, since $\Phi^a \Phi_a$ does not depend on the angular coordinate itself, we see that there exist *closed time-like curves* in the neighbourhood of the singularity of the Kerr solution.

22.4   Kerr-Schild coordinates

Another system of coordinates in which it is useful to present the Kerr solution can be seen by making the transformation

$$x + iy = (R - ia)e^{i\Phi}\sin\Theta$$

$$z = R\cos\Theta$$

$$\tau = T. \qquad (22.4.1)$$

Then the metric becomes

$$ds^2 = d\tau^2 - dx^2 - dy^2 - dz^2$$
$$- \frac{2MR^3}{R^4 + a^2 z^2}\left[\frac{R(x\,dx + y\,dy) + a(x\,dy - y\,dx)}{R^2 + a^2} + \frac{z\,dz}{R} + d\tau\right]^2, \qquad (22.4.2)$$

where the function $R(x, y, z)$ is determined implicitly by

$$R^4 - (x^2 + y^2 + z^2 - a^2)R^2 - a^2 z^2 = 0. \qquad (22.4.3)$$

These coordinates are called 'quasi-Minkowskian' on account of the way the flat 'background' metric $ds^2 = d\tau^2 - dx^2 - dy^2 - dz^2$ is clearly exhibited. And it is convenient to study properties of the other coordinate systems by viewing the surfaces of constant R, Θ, Φ in an artificial Euclidean 3-space with coordinates x, y, z. In particular the structure of the disk at $R = 0$ becomes more apparent in this system. For at $R = 0$ we have $z = 0$ with

$$x = -a \sin \Phi \sin \Theta \qquad (22.4.4)$$
$$y = a \cos \Phi \sin \Theta,$$

where Φ runs from 0 to 2π and Θ runs from 0 to $\pi/2$. Thus $\Theta = 0$ corresponds to the centre of the disk, whereas $\Theta = \pi/2$ corresponds to its boundary (the ring).

The surfaces of constant R for $R \neq 0$, on the other hand, form a family of confocal ellipsoids: this may be made slightly more evident if we rewrite (22.4.3) in the form

$$\frac{x^2 + y^2}{R^2 + a^2} + \frac{z^2}{R^2} = 1. \qquad (22.4.5)$$

From (22.4.1) we can deduce a similar equation giving the surface of constant Θ, viz.:

$$\frac{x^2 + y^2}{a^2 \sin^2 \Theta} - \frac{z^2}{a^2 \cos^2 \Theta} = 1. \qquad (22.4.6)$$

But here we recognise these surfaces as a family of hyperboloids, asymptotic to the cones defined by

$$(x^2 + y^2)^{1/2} = z \tan \theta.$$

In each case if $\cos \Theta$ is positive then we only take that portion of the hyperboloid (or cone) for which z is positive; and if $\cos \Theta$ is negative we restrict ourselves to negative values of z.

From the way in which the Kerr-Schild metric is written it should be evident that it is of the form

$$g_{ab} = \eta_{ab} + h\ell_a \ell_b \qquad (22.4.7)$$

where η_{ab} is a flat metric and ℓ_a is a vector which can be shown to be *null*.

In fact, the vector ℓ_a has an important role in the geometry of the Kerr solution: it is a *principal null vector* of the Riemann tensor, i.e. it satisfies

$$\ell^c \ell_{[a} R_{b]cd[e} \ell_{f]} \ell^d = 0. \qquad (22.4.8)$$

For a generic vacuum space-time the Riemann tensor admits four independent solutions of this relation; whereas for the Kerr solution they degenerate in pairs into just two independent solutions. Thus the Kerr solution is said to be a type (2,2) solution, or type D (for degenerate).

Note that in the Kerr-Schild form the principal null direction ℓ_a is exhibited explicitly, but the other one (which we shall call n_a) is not so evident.

The situation can be remedied to some extent by reverting to Boyer-Lindquist coordinates. There the two principal null directions are on an equal footing with one another, apart from a change of sign in the radial component. Thus in the (t, r, θ, ϕ) system we have

$$\ell^a = [(r^2 + a^2)/\Delta, -1, 0, a/\Delta],$$
$$n^a = [(r^2 + a^2)/\Delta, 1, 0, a/\Delta];$$
(22.4.9)

whereas in the Kerr-Eddington system (T, R, Θ, Φ) we have

$$\ell^a = [0, -1, 0, 0]$$
$$n^a = [(R^2 + a^2)/\Delta, 1, 0, a/\Delta].$$
(22.4.10)

Note that in the Kerr-Eddington system the principal direction ℓ^a assumes a particularly simple form. This, ultimately, is the reason why this form of the metric is so useful for many calculations.

Further transformations exist which in this respect reverse the roles of ℓ^a and n^a. To pursue the matter further here would be beyond our brief, but we note the importance of such considerations and refer the interested student to the original literature.

22.5   Energy extraction

In the case of the Schwarzschild solution the horizon or stationary null surface that acts as a 'one-way membrane', leading to the black hole characteristics of that field, coincides with the 'infinite red-shift' surface, the surface where the T^a Killing vector goes null. This is of course at $r = 2M$.

In the case of the Kerr solution these surfaces are generally *distinct*. The one-way membrane (for $a \leq M$) is located at

$$r_+ = M + (M^2 - a^2)^{1/2},$$

which is not surprisingly also the location of the spurious pseudo-singularity in the Boyer-Lindquist system. There is also a second, *inner* horizon given by $r_- = M - (M^2 - a^2)^{1/2}$ that is probably of less significance in physical considerations. But what is perhaps remarkable is the infinite red-shift surface

$$r_0 = M + (M^2 - a^2 \cos^2 \theta)^{1/2}$$

which lies *outside* of the horizon, except at the poles where they touch.

Thus for $a \leq M$ there is an intermediate region, outside the horizon, but inside the infinite red-shift surface. This is called the 'ergosphere'. Associated with this region are a host of important considerations leading to striking physical consequences. For example, a particle with mass m_0 coming from far away with total energy E_0 and positive angular momentum Φ_0 can enter the ergosphere and decay into a pair of particles. One particle, with mass m_1, negative energy E_1, and negative angular momentum Φ_1, falls through the one-way membrane and is lost to the black hole. The other particle, with mass m_2, positive energy E_2, and positive angular momentum Φ_2, leaves the ergosphere and returns to the depths of space. The extraordinary feature of this so-called 'Penrose process' is that the energy E_2 of the particle emitted back to infinity is *greater* than the energy E_0 of the original incoming particle. Some of the energy of the black hole has thus been 'extracted'.

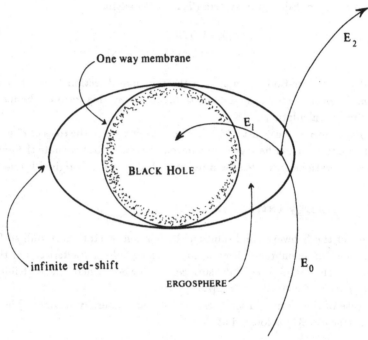

Figure 22.1. Energy extraction from a Kerr black hole. The outgoing energy E_2 is greater than the incoming energy E_0. A particle with negative energy E_1 is lost down the black hole.

The details are complex, but the reasoning behind this phenomenon is simple enough. Far away from the black hole the energy of the incoming particle, as measured by a stationary observer, is $E_0 = m_0 T^a \xi_a$, where ξ_a is the unit tangent to the geodesic trajectory of the particle. Since $T^a \xi_a$ is conserved along the particle's trajectory, we have

$$m_0 T^a \xi_a = m_1 T^a \eta_a + m_2 T^a \zeta_a$$

at the point of decay, where η_a and ζ_a are respectively the tangents to the first and second decay produces. But since T^a is *space-like* at the point of decay, we can arrange that $T^a \eta_a < 0$ (even though the unit vector η_a is time-like and future-pointing). Thus $E_1 = m_a T^a \eta_a$ is negative, and $E_2 = m_2 T^a \zeta_a$ will be measured at infinity to be greater than the original E_0.

From the Penrose process we are led on towards a host of fascinating ideas and problems—black hole thermodynamics, the area theorem, Hawking radiation, black hole explosions, and so on; and more deeply and generally to questions of that most elusive of twentieth century sciences, *quantum gravity*. How is the Hilbert space structure of quantum theory, with its system of observables as linear operators, and its subtle probabilistic interpretation, to be correctly wedded to the quintessentially classical Riemannian geometry of Einstein's theory? It is beyond our scope to attempt even briefly to address this question here; but suffice is to say that the answer remains unknown: as Einstein remarked long ago in 1954, '*Only a significant progress in the mathematical methods can help here. At the present time the opinion prevails that a field theory must first, by "quantization", be transformed into a statistical theory of field probabilities according to more or less well-established rules. I see in this method only an attempt to describe relationships of an essentially nonlinear character by linear methods.*'

Exercises for chapter 22

[22.1] Verify the equivalence of (22.1.1) and (22.1.4), and that as $a \to 0$ both reduce to (22.1.5).

[22.2] Integrate (22.2.1) so as to obtain explicit formulae for the transition from Boyer-Lindquist coordinates to Kerr-Eddington coordinates.

[22.3] For which values of (r, θ) is $T^a T_a$ negative? For which values of (r, θ) is $\Phi^a \Phi_a$ positive? Find expressions for r and θ in terms of $T^a T_a$, $T^a \Phi_a$, and $\Phi^a \Phi_a$.

[22.4] Verify the transition from (22.2.3) to (22.4.2) via the relations (22.4.1).

[22.5] Let h_{ij} be a metric tensor on a region of space-time and l^i a vector field which is null with respect to h_{ij}. A new metric g_{ij} is defined by

$$g_{ij} = h_{ij} + l_i l_j$$

where $l_i = h_{ij} l^j$. Prove that the inverse of g_{ij} is given by

$$g^{ij} = h^{ij} - l^i l^j$$

where h^{ij} is the inverse of h_{ij}. Let ∇_i be the metric connection defined such that $\nabla_i h_{jk} = 0$. A new connection $\tilde{\nabla}_i$ is defined according to the scheme

$$\tilde{\nabla}_i \xi_j = \nabla_i \xi_j - Q_{ij}^k \xi_k$$

for all smooth ξ_i. Show that if $\tilde{\nabla}_i g_{jk} = 0$ then

$$2Q_{ij}^n = \nabla_i(l_j l^n) + \nabla_j(l_i l^n) + l^n l^k \nabla_k(l_i l_j) - \nabla^n(l_i l_j).$$

Hence show that Q_{ij}^n satisfies the following relations:

$$Q_{in}^n = 0,$$
$$l_n Q_{ij}^n = -\frac{1}{2}(l_i A_j + l_j A_i),$$
$$l^i Q_{ij}^n = \frac{1}{2}(l^n A_j + l_j A^n),$$
$$l^i A_i = 0,$$

where $A_i = l^k \nabla_k l_i$. If R_{ij} is the Ricci tensor associated with h_{ij}, then the Ricci tensor associated with g_{ij} is given by

$$\frac{1}{2}\tilde{R}_{ij} = \frac{1}{2}R_{ij} + \nabla_{[i}Q_{m]j}^m + Q_{j[i}^n Q_{m]n}^m.$$

Suppose that h_{ij} is flat. Show that $\tilde{R}_{ij}l^i l^l = 0$ if and only if l^i satisfies the geodesic equation

$$l^i \nabla_i l^j = \Phi l^j$$

for some scalar Φ.

[22.6] Show that an anti-symmetric tensor Y_{ab} satisfies the Yano equation

$$\nabla_{(a}Y_{b)c} = 0$$

if and only if its dual $^*Y_{ab}$ (see § 3.5 for the definition of duality) satisfies

$$\nabla_c {}^*Y_{ab} = -2\xi_{[a}g_{b]c}$$

for some vector ξ_a. If Y_{ab} is a Yano tensor show that $K_{ab} = Y_{am}Y^m{}_b$ satisfies $\nabla_{(a}K_{bc)} = 0$. Show that

$$Y_{m(a}R_{b)}^m = 0$$

and that

$$^*Y_{m(a}R_{b)}^m = 2\nabla_{(a}\xi_{b)}$$

where R_{ab} is the Ricci tensor, and hence that for a vacuum space-time that admits a Yano tensor (as does the *Kerr solution*) the associated vector ξ_a is a *Killing vector*. Show that if ξ^a is a Killing vector then so is $\eta^a = K^{ab}\xi_b$.

[22.7] Show that the horizon $r = r_+$ is a null surface.

23   Homogeneous and isotropic three-spaces

Cést une sphère dont le centre est partout, la circonférence nulle part.

—Pascal (**Les Pensées**, 348)

AS AN INVITATION to the discussion of cosmology in later chapters we ask: which three-dimensional metrics are *homogeneous* and *isotropic*? Precise definitions of these terms will be given later in chapter 26 where we develop more substantial mathematical machinery. In this chapter we shall get along by appealing to intuition. For the moment we think of 'homogeneous' as meaning 'the same at every point' and 'isotropic about a point p' as meaning that all directions at p are equivalent.

In particular, 'isotropic at p' must entail 'spherically symmetric about p'. Further, homogeneity and isotropy about a point taken together must entail that there are no preferred vectors at *any* point, i.e. isotropy about *every* point.

In a three-dimensional Riemannian manifold, the Riemann tensor can be expressed in terms of the Ricci tensor, Ricci scalar and metric tensor via the relation

$$R_{abcd} = R_{ac}g_{bd} - R_{bc}g_{ad} + R_{bd}g_{ac} - R_{ad}g_{bc} - \frac{1}{2}R(g_{ac}g_{bd} - g_{bc}g_{ad}). \tag{23.1}$$

Thus any anisotropy in the Riemann tensor must in this sense arise directly from the Ricci tensor. By raising an index on the Ricci tensor with the metric we may define a linear transformation on vectors at any point by $\ell^a \rightarrow R_b^a \ell^b$. This transformation will have eigenvectors, which are therefore preferred directions, unless it is proportional to the identity. Thus isotropy in three dimensions implies $R_{ab} = K g_{ab}$ for some K. Taking the trace, we see that this means

$$R_{ab} = \frac{1}{3}Rg_{ab},$$

whence by (23.1) we obtain

$$R_{abcd} = \frac{1}{6}R(g_{ac}g_{bd} - g_{ad}g_{bc}). \tag{23.2}$$

The Bianchi identity (see section 6.4) applied to (23.2) then forces the Ricci scalar R to be *constant*. (This could have been anticipated, since otherwise $\nabla_a R$ would be a preferred vector!) Thus the requirement of isotropy about every point leads to a metric satisfying (23.2) with constant R. These spaces are referred to as *spaces of constant curvature*.

If the metric g_{ab} has constant curvature then clearly so has the metric $\hat{g}_{ab}$ defined by

$$\hat{g}_{ab} = \lambda^2 g_{ab} \tag{23.3}$$

for constant λ. It is a straightforward exercise to verify that the curvature tensors associated with these metrics satisfy $\hat{R}_{bcd}{}^a = R_{bcd}{}^a$ and thus

$$\hat{R} = \lambda^{-2} R. \tag{23.4}$$

This will change the magnitude of the constant R, but not its sign, so there are only *three essentially distinct cases* to consider. Introducing the parameter k by $R = 3k$ we may suppose that k is $0, 1$ or -1.

Since an isotropic metric is a *fortiori* spherically symmetric, we may make an ansatz as in chapter 15, i.e. $ds^2 = A^2(r)dr^2 + B^2(r)(d\theta^2 + \sin^2\theta d\phi^2)$, and then redefine the r-coordinate so that

$$ds^2 = d\chi^2 + f^2(\chi)(d\theta^2 + \sin^2\theta d\phi^2). \tag{23.5}$$

A straightforward calculation (exercise) now leads to

$$R = -(2\frac{f''}{f} - \frac{1}{f^2} + \frac{(f')^2}{f^2}). \tag{23.6}$$

To integrate this we have

$$[f(f')^2]' = f^2 f'(2\frac{f''}{f} + \frac{(f')^2}{f^2}) = -Rf^2 f' + f'$$

so that $f(f')^2 = f - \frac{1}{3}Rf^3$ where we have set the constant of integration equal to zero. Now with $R = 3k$ the solutions are

- i) $k = 0$ $f = \chi$
- ii) $k = 1$ $f = \sin\chi$
- iii) $k = -1$ $f = \sinh\chi$.

As we might have expected, the case $k = 0$, with zero curvature, is flat space in polar coordinates. For $k = 1$ the metric is

$$ds^2 = d\chi^2 + \sin^2\chi(d\theta^2 + \sin^2\theta d\phi^2). \tag{23.7}$$

This is the metric of the three-sphere S^3, i.e. it is the metric induced on the set of points unit distance from the origin in a four-dimensional space with a Euclidean metric. To see this, suppose the sphere is given by $X^2 + Y^2 + Z^2 + W^2 = 1$ in the space with metric

$$ds^2 = dX^2 + dY^2 + dZ^2 + dW^2. \tag{23.8}$$

We may introduce polar coordinates

$$X = R\sin\theta\cos\phi\sin\chi$$
$$Y = R\sin\theta\sin\phi\sin\chi$$
$$Z = R\cos\phi\sin\chi$$
$$W = R\cos\chi$$

so that the three sphere is just $R = 1$, and the metric (23.8) becomes

$$ds^2 = dR^2 + R^2(d\chi^2 + \sin^2 \chi(d\theta^2 + \sin^2 \theta d\phi^2)). \tag{23.9}$$

Now, as claimed, (23.9) reduces to (23.7) on $R = 1$. For $k = -1$, the metric is

$$ds^2 = d\chi^2 + \sinh^2 \chi(d\theta^2 + \sin \theta^2 d\phi^2). \tag{23.10}$$

This is the metric induced on the hyperboloid of unit time-like vectors (future pointing) in Minkowski space. To see this, suppose the hyperboloid is given by

$$T^2 - X^2 - Y^2 - Z^2 = 1, \quad T > 0 \tag{23.11}$$

in the space with metric

$$ds^2 = dT^2 - dX^2 - dY^2 - dZ^2. \tag{23.12}$$

Proceeding by analogy with S^3 we introduce coordinates

$$X = R \sin \theta \cos \phi \sinh \chi$$
$$Y = R \sin \theta \sin \phi \sinh \chi$$
$$Z = R \cos \theta \sinh \chi$$
$$T = R \cosh \chi.$$

The hyperboloid is $R = 1$ and the Minkowski metric is

$$ds^2 = dR^2 - R^2(d\chi^2 + \sinh^2 \chi(d\theta^2 + \sin^2 \theta d\phi^2))$$

which reduces to (23.10) on $R = 1$ except for the change of sign, which is just conventional.

The requirements of constant curvature and isotropy about one point have led us to three metrics. To verify that these three metrics are in fact homogeneous and isotropic we must be more precise about these terms. The appropriate definitions are in terms of the symmetries of the spaces, in other words, in terms of isometry groups (cf. chapter 26). By *isotropy* about a point we mean that the stabilizer of that point (i.e. the group of isometries that fix the point) is isomorphic to the three-dimensional rotation group $SO(3)$. By *homogeneity* we mean that any point can be taken to any other point by an isometry, i.e. that the isometry group is transitive. Since the spaces under consideration are three dimensional, this means that the isometry group must be at least six dimensional. However, we know that in dimension n, the largest possible isometry group has dimension $\frac{1}{2}n(n + 1)$ (see exercise 15.3). This is *six* in dimension three so that a homogeneous and isotropic space has the largest possible isometry group and is therefore also referred to as a *maximally symmetric* space.

We must consider the isometry groups of the three metrics that are under consideration here. For $k = 0$ and flat space the isometry group is the semi-direct product of three translations and the rotation group $SO(3)$. Evidently, this is transitive and the stabilizer of each point is $SO(3)$. Flat space is homogeneous and isotropic! For $k = 1$

the isometry group of S^3 must certainly include all of the four dimensional rotation group $SO(4)$. However this is already six-dimensional and (apart possibly from some discrete isometries) must be the whole isometry group. Clearly there is a rotation taking any point of S^3 to any other point, so the group is transitive. Clearly also the subgroup fixing any point of S^3 is the subgroup of $SO(4)$ fixing one direction, which is the three dimensional rotation group $SO(3)$.

For the hyperboloid the isometry group is the group of Lorentz transformations, written $SO(1,3)$. The statement that this is transitive is simply the statement that there is a Lorentz transformation between any two observers. Again the stabilizer of a point is the subgroup of $SO(1,3)$ that fixes a time-like direction, which is clearly $SO(3)$. This completes the verification that the metrics (23.7), (23.10) and flat space together with the rescaling (23.3) give all homogeneous and isotropic three-spaces.

Exercises for chapter 23

[23.1] Prove formula (23.1). (Hint: define $S_{ab} = \varepsilon_a{}^{cd}\varepsilon_b{}^{ef}R_{cdef}$ and attempt to express R_{abcd} in terms of S_{ab}.)

[23.2] Check that the Ricci scalar of the metric (23.5) is given by (23.6).

[23.3] Show that the three basic metrics of constant curvature can be written in the equivalent form

$$ds^2 = \frac{dr^2}{1 - kr^2} + r^2(d\theta^2 + \sin^2\theta d\phi^2) \quad k = 0, 1, -1.$$

[23.4] Show that if a space-time is empty and isotropic it must be Minkowski space. (Hint: exercise 17.7.)

[23.5] Show that the total volume of the 3-space is *infinite* for $k = -1$ and *finite* $(2\pi^2)$ for $k = +1$.

24   Cosmology: kinematics.

THE AIM OVER the next two chapters is to construct a solution of Einstein's equations with sources that will provide a model for the   large scale features of the universe. First, we must find a reasonable form for the metric and energy-momentum tensor consistent with the observed symmetries of the universe. Then we shall be led to specific cosmological models by the imposition of the Einstein equations.

We are thinking of the *average* features of the universe on the scale of tens of millions of light years and we may regard the basic building blocks as clusters of galaxies. The first observational fact about the universe that we must use is that the observed distribution of the clusters of galaxies is *isotropic* to a high degree. If we assume that our position is in no particular way privileged, we must assume the universe is isotropic about every point, which leads to an assumption of *homogeneity*.

We must distinguish a preferred class of observers, namely those that actually *see* the universe as isotropic. Thus our cosmological model admits a *preferred* time-like vector field u^a, tangent to the world lines of the preferred or 'fundamental' observers.

We next assume that there is a preferred family of hypersurfaces orthogonal to the world lines of the fundamental observers (i.e. that the vector field u^a is hypersurface-orthogonal). These hypersurfaces are the instantaneous rest spaces of the fundamental observers. We incorporate the observation of isotropy and the consequent assumption of homogeneity into the cosmological model by requiring that these hypersurfaces should be homogeneous and isotropic three-spaces in the sense of chapter 23.

To write down the metric of the model we must first introduce appropriate coordinates. We label the instantaneous rest spaces by proper-time t along the fundamental observers. We choose the three-space coordinates so that they are constant along the world lines of the fundamental observers. These are referred to as *comoving coordinates*.

By assumption, the metric of the constant t hypersurface is of the form of (23.3) where the factor λ is allowed to be a function of t, and g_{ab} is one of the three possibilities found in chapter 23. Furthermore there can be no $dt dx^i$ cross terms in the full four-dimensional metric because the world lines of the fundamental observers are orthogonal to the constant t hypersurfaces. Finally, because t is proper time, the coefficient of the dt^2 term in the metric is unity. The metric consistent with these assumptions is therefore

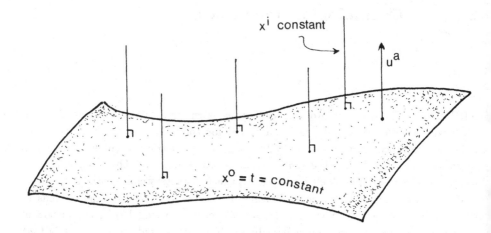

Figure 24.1. Comoving coordinates.

$$ds^2 = dt^2 - R^2(t)g_{ij}dx^i dx^j, \tag{24.1}$$

where

$$g_{ij}dx^i dx^j = \begin{cases} d\chi^2 + \chi^2(d\theta^2 + \sin^2\theta d\phi^2) & \text{for } k = 0 \\ d\chi^2 + \sin^2\chi(d\theta^2 + \sin^2\theta d\phi^2) & \text{for } k = 1 \\ d\chi^2 + \sinh^2\chi(d\theta^2 + \sin^2\theta d\phi^2), & \text{for } k = -1 \end{cases},$$

or equivalently

$$g_{ij}dx^i dx^j = \frac{dr^2}{1 - kr^2} + r^2(d\theta^2 + \sin^2\theta d\phi^2)$$

by exercise [23.3] where the scale factor λ of equation (23.3) is denoted $R(t)$. This metric form is known as the *Friedmann-Robertson-Walker* metric. For later use, we remark that the fundamental velocity u^a is just the gradient of t: in these coordinates

$$u_a = \nabla_a t = (1, 0, 0, 0). \tag{24.2}$$

To see the significance of the scale factor we consider two fundamental observers P and Q that intersect a particular constant t slice, say $t = t_0$, at p_0 and q_0. The proper distance between p_0 and q_0 will be $R(t_0)s_0$ where s_0 is the proper distance measured in the metric $g_{ij}dx^i dx^j$. At a later time t_1 since P and Q are fundamental observers and are therefore at constant values of x^i, the proper distance will be $R(t_1)s_0$. That is, the distance will change only by virtue of the scale factor $R(t)$ changing. The scale factor $R(t)$ therefore encodes the *relative motion* of the fundamental observers.

We now need to construct an energy momentum tensor T_{ab} consistent with our assumptions, and with our notion of the matter content of the universe. It can be made only from the velocity u^a, the metric g_{ab} and functions of t since any other ingredient would violate the assumed isotropy and homogeneity. The only possibility is therefore $T_{ab} = A(t)u_a u_b + B(t)g_{ab}$ for two functions A and B which we write, re-labelling the two functions, as

$$T_{ab} = (\rho + p)u_a u_b - pg_{ab}. \tag{24.3}$$

Therefore the *symmetry alone* demands that T_{ab} have the form of a *perfect fluid* stress tensor, with the pressure p and density ρ being functions only of t. The fluid four-velocity is u^a, the velocity of the fundamental observers—so we have a consistent picture: the fundamental observers are those at rest in the fluid. We may think of the 'fluid' in the first instance as being a smoothed out 'gas' of clusters of galaxies.

We obtain one relation linking the matter content and the metric from the conservation equation applied to (24.3):

$$\nabla_a T^{ab} = 0 = u^b \nabla_a((\rho + p)u^a) + (\rho + p)u^a \nabla_a u^b - \nabla^b p. \tag{24.4}$$

Here p is a function only of t, and u^a is a unit vector; so this equation splits into two, parallel to u^a and orthogonal to u^a (cf. section 3.6):

$$u^a \nabla_a u^b = 0, \tag{24.5}$$

$$\nabla_a((\rho + p)u^a) = u^b \nabla_b p. \tag{24.6}$$

The first of these is trivially satisfied:

$$u^a \nabla_a u_b = u^a \nabla_b u_a \quad \text{(by 24.2)}$$
$$= \frac{1}{2} \nabla_b(u^a u_a)$$
$$= 0,$$

since u^a is a unit vector. For the second we must compute the divergence of u^a:

$$\nabla_a u^a = \nabla_a \nabla^a t \quad \text{(by 24.2)}$$
$$= \frac{1}{\sqrt{-g}}(\sqrt{-g}g^{ab}t_{,a})_{,b} \quad \text{(by exercise 7.10)}$$
$$= \frac{3\dot{R}}{R} \quad \text{(by substitution from 24.1)}$$

where we use a dot for differentation with respect to t. Now (24.6) becomes

$$\dot{\rho} + \frac{3\dot{R}}{R}(\rho + p) = 0. \tag{24.7}$$

This in turn can be written in two ways that are useful later:

$$\frac{d}{dR}(\rho R^3) = -3pR^2 \tag{24.8}$$

or

$$(\rho R^3)^\bullet = -p(R^3)^\bullet \tag{24.9}$$

In the second form it is readily apparent that this equation expresses a sort of relativistic *conservation of energy*:

$$\frac{dM}{dt} = -p\frac{dV}{dt} \tag{24.10}$$

where M denotes the mass ρV contained within a comoving volume V. From our discussion of the scale factor $R(t)$, clearly V is of the form $R^3(t) \times V_0$ where V_0 is measured by g_{ij} and so is independent of time. Now multiplication of (24.9) by V_0 leads to (24.10).

We shall obtain more relations between the matter content and scale factor by imposing the Einstein equations. Before that, and to discuss *observations* in cosmological models, we must consider *the propagation of light* or equivalently the *null geodesic equation*. We suppose the metric universe is given by (24.1) and that we as observers are situated at the origin of the spatial coordinates. Then the only null geodesics reaching or leaving us are radial and the appropriate Lagrangian is

$$L = \dot{t}^2 - \frac{R^2(t)\dot{r}^2}{1 - kr^2} = 0.$$

The incoming $(-)$ and outgoing $(+)$ null geodesics therefore satisfy

$$\int \frac{dt}{R(t)} = \pm \int \frac{dr}{(1 - kr^2)^{\frac{1}{2}}}. \tag{24.11}$$

The first consequence of (24.11) is the *cosmological red-shift* (since the fundamental observers are in relative motion we should expect some red-shift or time dilation effect between them): suppose a galaxy p at coordinate r_1 releases a flash of light at t_1 that is received by us at $r = 0$ at t_0, as in figure 24.2. By virtue of (24.11) we have

$$\int_0^{r_1} \frac{dr}{(1 - kr^2)^{\frac{1}{2}}} = \int_{t_1}^{t_0} \frac{dt}{R(t)} \tag{24.12}$$

and for a later flash

$$= \int_{t_1+\delta t_1}^{t_0+\delta t_0} \frac{dt}{R(t)}.$$

Thus

$$\frac{\delta t_1}{R(t_1)} = \frac{\delta t_0}{R(t_0)}.$$

This is a cosmological time dilation and leads to a cosmological red-shift:

$$\frac{\lambda_0}{\lambda_1} = \frac{\lambda_1 + \delta\lambda}{\lambda_1} = \frac{R(t_0)}{R(t_1)}$$

where λ_1 is the emitted wave-length and λ_0 the received wave-length. The red-shift z is defined by

$$z = \frac{\delta\lambda}{\lambda} = \frac{R(t_0) - R(t_1)}{R(t_1)} \tag{24.13}$$

and is *positive* for an expanding universe.

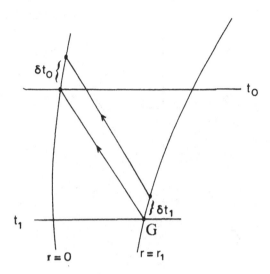

Figure 24.2. The cosmological red-shift. A galaxy G at r_1 releases a flash of light at time t_1 that is received by us at $r = 0$ at time t_0. By the time the light is received it has been red-shifted.

The crucial importance of the red-shift lies in the fact that spectroscopic observations are sufficiently precise that wavelengths can be measured very accurately.

From (24.12) we see that the past light cone of the origin at time t_0, i.e the set of events from which null geodesics arrive at the origin at time t_0, is the surface $r_1 = f(t_1)$ defined by the equation

$$\int_0^{r_1} \frac{dr}{(1 - kr^2)^{\frac{1}{2}}} = \int_{t_1}^{t_0} \frac{dt}{R(t)}. \tag{24.14}$$

Similarly, for the future light cone of the origin at time t_1 we obtain the same equation.

Events *beyond* these surfaces have never been seen by us and will never see us, respectively.

Now, if it should happen that the right hand side of (24.14) has a finite limit as t_0 recedes into the future then there will be events which are never visible to the fundamental observer at the origin. There will therefore be a *cosmological event horizon* associated with that observer. Note the similarities with the event horizon of the Schwarzschild solution (cf. chapter 19): this is a *null surface*, so once crossed by an object on a time-like path it can never be recrossed; however there is nothing special at the moment of crossing for the crosser. The difference is that the cosmological event horizons of different fundamental observers are different, while in the Schwarzschild solution the event horizon is the same for all external observers.

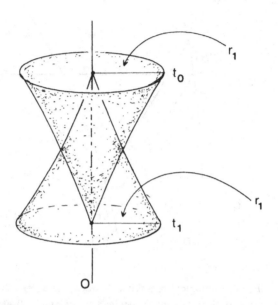

Figure 24.3. The future light cone of the origin at time t_1 reaches the same coordinate radius r_1 as the past light cone of the origin at time t_0.

Another possibility is that (24.14) could have a finite limit as the lower limit of integration recedes into the past, either to $-\infty$ or to some initial value. In this case there will be fundamental observers who neither have seen nor have been seen by the fundamental observer at the origin since the universe began (or since the infinite past). The limiting future light cone, that is the future light cone of the origin 'at the beginning', is referred to as a *particle horizon*.

One main object of observations in cosmology must be to find the form of the scale function $R(t)$. This is involved in the equation (24.14) for the light cone but one has no way of making direct observations of r_1 which is related to proper distance in the metric g_{ij}. What is needed is some sort of observable definition of distance. One such notion is the *apparent area distance* d_A. This is defined as follows.

The past light cone of the observer at t_0 meets the surface t_1 in a sphere of coordinate radius r_1. Its area is therefore $4\pi r_1^2 R^2(t_1)$. We imagine an object, say a galaxy, covering an area A on this sphere, and subtending a solid angle $\delta\omega$ at the earth. In flat space we would have $A = r^2 \delta\omega$, so in curved space we define the *apparent area distance* d_A by this equation:

$$(d_A)^2 = \frac{A}{\delta\omega}. \tag{24.15}$$

We also have that

$$\frac{\delta\omega}{4\pi} = \frac{A}{4\pi r_1^2 R^2(t_1)}$$

so that

$$d_A = r_1 R(t_1). \tag{24.16}$$

In principle then if there are objects in the universe whose *actual* size we know then we can measure directly their apparent area distance by (24.15) and obtain this as a function of their red-shift, which also is directly observable. Then (24.13), (24.14) and (24.16) can in principle be used to turn d_A as a function of z into R as a function of t.

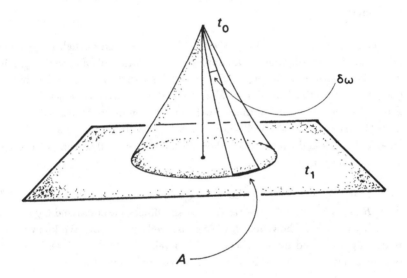

Figure 24.4. The geometrical arrangement for the definition of apparent area distance.

In practice of course things are not so easy and various other operational definitions of distance must be used. Different definitions are useful over different scales and a consistent set must be constructed.

Exercises for chapter 24

[24.1] Verify that (24.14) does indeed give the future light cone of the origin at time t_1.

[24.2] What kinds of horizons are associated with the $k = 0$ FRW metric and the following scale factors:
   (i) $R(t) = \cosh t \quad -\infty < t < \infty$,
   (ii) $R(t) = e^t \quad -\infty < t < \infty$,
   (iii) $R(t) = t^{1/2} \quad t > 0$.

[24.3] Show that if $k = 0$ and $R(t) = t^{2/3}$ then the area distance has a maximum, so that objects at large distances have apparent areas that increase with increasing remoteness.

[24.4] *Geometrical optics*: we aim to show that light with high enough frequency follows null geodesics by finding an approximate solution of Maxwell's equations. Take the potential to be $A_a = B_a exp(iS/\varepsilon)$ so that small ε means high frequency. Define the field $F_{ab} = 2\nabla_{[a}A_{b]}$ and for convenience set $k_a = \nabla_a S$. From the $O(\varepsilon^{-1})$ term of the Lorentz gauge condition $\nabla^a A_a = 0$ and the $O(\varepsilon^{-2})$ term of the Maxwell equation $\nabla^a F_{ab} = 0$ deduce that [cf. exercise 14.4] $k_a B^a = 0$ and $k_a k^a = 0$ and hence that $k^b \nabla_b k^a = 0$ so that k^a is tangent to null geodesics.

[24.5] Note that the only part of the metric relevant to the light cones is $ds^2 = dt^2 - R^2(t)d\chi^2$. Let ℓ^a be the tangent to an affinely parametrized light ray and let $\chi^a \nabla_a = \partial/\partial\chi$. Show that $\ell_a \chi^a$ is conserved on the ray. We know that the frequency measured by an observer with 4-velocity U^a is $\ell_a U^a$ and that ℓ^a is, of course, null. Use these facts to rederive the red-shift formula.

[24.6] Suppose that a star is known to have a luminous power output L, while we detect an energy flux per unit area ℓ. A sensible measure of its distance away would then appear to be

$$d_L = \sqrt{\frac{L}{4\pi\ell}}.$$

Show that this measure is related to the one in the text by

$$d_L = (1 + z)^2 d_A.$$

25 Cosmology: dynamics

THE DYNAMICAL RELATION between the geometry represented by the scale factor R and the matter represented by the density ρ and pressure p are determined by the Einstein equations. We must calculate the Ricci tensor of the Friedmann-Robertson-Walker metric (24.1) and substitute into the Einstein equations with the energy-momentum tensor (24.3):

$$R_{ab} = -8\pi G(T_{ab} - \frac{1}{2}T g_{ab}).$$

Just two equations result from this:

$$R_{00} : 3R^{-1}\ddot{R} = -4\pi G(\rho + 3p) \tag{25.1}$$

$$R_{ij} : R^{-1}\ddot{R} + 2R^{-2}\dot{R}^2 + 2kR^{-2} = 4\pi G(\rho - p). \tag{25.2}$$

If we eliminate $\ddot{R}$ between these we obtain

$$\dot{R}^2 + k = \frac{8}{3}\pi G \rho R^2. \tag{25.3}$$

This is known as the *Friedmann equation* and has the character of an energy equation in Newtonian mechanics, as we see by writing it in the form

$$\frac{1}{2}\dot{R}^2 - G(\frac{4}{3}\pi R^3 \rho)R^{-1} = \text{constant} = -\frac{1}{2}k.$$

The conservation equation (24.7) must be a consequence of (25.1) and (25.2), so if we include it written as:

$$\dot{\rho} + 3R^{-1}\dot{R}(\rho + p) = 0 \tag{25.4}$$

then the system (25.3) + (25.4) is equivalent to (25.1) + (25.2). These are the dynamical equations determining the Friedmann-Robertson-Walker cosmological models. Since there are three unknowns, namely R, ρ and p, we evidently need a third equation to close the system. What is needed is an *equation of state* relating p and ρ. This will reflect our knowledge about the details of the constitution of the 'gas' or fluid of which the universe is composed and must be an *independent* assumption. This is a situation much as we have already discussed in the context of special relativity in section 3.6.

As a first example, we shall consider the assumption that the matter filling the universe is *dust* or 'incoherent' matter, i.e. that the *pressure vanishes*. From the conservation equation in the form (24.8):

$$\frac{d}{dR}(\rho R^3) = 0$$

we deduce that

$$\rho = MR^{-3} \tag{25.5}$$

for some constant M. The Friedmann equation becomes

$$\dot{R}^2 + k = \frac{1}{3}\kappa M R^{-1} \tag{25.6}$$

where we set $8\pi G = \kappa$ temporarily. The solution depends on k and we shall discuss the three cases separately:

Case (i). $k = 0$: $R\dot{R}^2 = \frac{1}{3}\kappa M$ so

$$R = (a(t - t_0))^{2/3} \tag{25.7}$$

where

$$a^2 = \frac{3}{4}\kappa M$$

Thus by (25.5) we obtain $\rho = \frac{4}{3}\kappa^{-1}(t - t_0)^{-2}$. The constant of integration is an arbitrary additive constant t_0 in t which we may set to zero.

For the two other cases, it is convenient to introduce a new independent variable ψ by

$$\frac{dt}{d\psi} = R \tag{25.8}$$

With this variable, (25.6) is

$$(\frac{dR}{d\psi})^2 = R^2 \dot{R}^2 = \frac{1}{3}\kappa M R - kR^2$$

whence also

$$\frac{d^2 R}{d\psi^2} = \frac{1}{6}\kappa M - kR.$$

Case (ii). $k = 1$:

$$\left.\begin{array}{l} R = \dfrac{1}{6}\kappa M(1 - \cos(\psi - \psi_0)) \\[2mm] t - t_0 = \dfrac{1}{6}\kappa M(\psi - \sin(\psi - \psi_0)) \end{array}\right\} \tag{25.9}$$

without loss of generality we may set t_0 and ψ_0 equal to zero. This is the parametric equation of a *cycloid*.

Case (iii). $k = -1$:

$$\left.\begin{array}{l} R = \dfrac{1}{6}\kappa M(\cosh\psi - 1) \\[3mm] t = \dfrac{1}{6}\kappa M(\sinh\psi - \psi) \end{array}\right\} \tag{25.10}$$

The three cases are shown graphically in figure (25.1).

All start from the origin so that initially the density is *infinite*. This is a *curvature singularity*. The $k = 1$ universe (the 'closed' universe) reaches a maximum expansion and then recontracts to another infinite density curvature singularity. The others (the 'open' universes) continue to expand indefinitely though at different rates: for $k = -1$ we see $R \sim t$ for large t while for $k = 0$, $R \sim t^{2/3}$. Apart from the discrete parameter k, there is one other free parameter M.

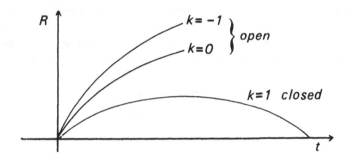

Figure 25.1. The three kinds of dust-filled universes

For another class of models, we consider the case of a universe filled uniformly with *electromagnetic radiation* (or with massless neutrinos). The equation state for this is $p = \frac{1}{3}\rho$ which with (24.8) gives

$$\rho = MR^{-4}. \tag{25.12}$$

The Friedmann equation is

$$\dot{R}^2 + k = \frac{1}{3}\kappa M R^{-4}$$

or, in terms of the variable ψ introduced by (25.8):

$$(\frac{dR}{d\psi})^2 = \frac{1}{3}\kappa M - kR^2.$$

The solutions this time are

$$(i) \ \ k = 0: \ \ R = (\frac{4}{3}\kappa M)^{\frac{1}{4}}t^{\frac{1}{2}} \qquad \text{(open)}$$

$$(ii) \ \ k = 1: \ \ R = (\frac{1}{3}\kappa M)^{\frac{1}{2}}\sin\psi \quad \text{(closed)}$$
$$t = (\frac{1}{3}\kappa M)^{\frac{1}{2}}(1 - \cos\psi)$$

$$(iii) \ \ k = -1: \ \ R = (\frac{1}{3}\kappa M)^{\frac{1}{2}}\sinh\psi \ \text{(open)}$$
$$t = (\frac{1}{3}\kappa M)^{\frac{1}{2}}(\cosh\psi - 1)$$

The picture for this case is qualitatively the same as figure (25.1). The differences are in the rate of expansion initially and finally and in the life-time of the closed universe.

All the models again have an initial curvature singularity. This is a general feature as may be seen by a consideration of equation (25.1). There we see that

$$\ddot{R} < 0 \quad \text{if} \quad (\rho + 3p) > 0.$$

Regardless of the equation of state we expect the second inequality to be satisfied by *all* reasonable matter, so the graph of R must be concave downwards. Thus if $\ddot{R}$ is ever positive (and it is measured as being positive now) then there must be a point in the past where $R = $ zero (see figure 25.2). In fact, the singularity must occur before a time

$$\tau = R_0 \dot{R}_0^{-1} \equiv H_0^{-1} \tag{25.13}$$

into the past.

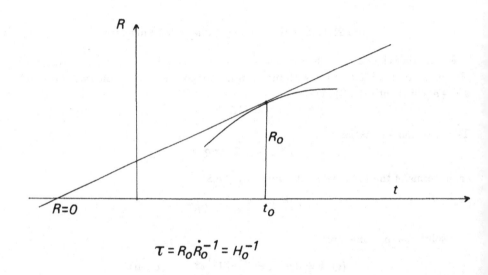

$$\tau = R_0 \dot{R}_0^{-1} = H_0^{-1}$$

Figure 25.2. The necessity for a cosmological singularity.

In summary, the picture that emerges from imposition of the Einstein equations on the Friedmann-Robertson-Walker metric is of a universe expanding away from an *initial singularity of infinite density*. The universe will expand forever if the space sections are infinite (open universe), or will recontract to another singularity if the space sections are finite (closed universe). Such features as the rate of expansion, the density and the life-time (if finite) depend on the particular model and equation of state.

Let us consider now how we may observationally distinguish between the different Friedmann-Robertson-Walker models. We define two parameters that we shall see are observationally determined. These are the *Hubble parameter* H with dimensions of inverse of time and the dimensionless *deceleration parameter* q:

$$H = \dot{R}R^{-1}; \quad q = -R\ddot{R}\dot{R}^{-2}. \tag{25.14}$$

For the rest of this chapter a *subscript zero* on a quantity denotes its value at the *present time*. We saw in (25.13) that H_0^{-1} gives an upper limit on the age of the universe. Now for k, we have from the Friedmann equation (25.3),

$$k = \frac{1}{3}\kappa\rho R^2 - \dot{R}^2 = \dot{R}^2(\frac{1}{3}\kappa H^{-2}\rho - 1). \tag{25.15}$$

The quantity

$$\rho_c = 3H^2\kappa^{-1} = \frac{3H^2}{8\pi G}$$

has the dimensions of density and is known as the 'critical' or 'magic' density. Its significance is that by equation (25.15) we have

$$k = \dot{R}^2(\frac{\rho}{\rho_c} - 1)$$

so that the sign of k depends on the ratio of the actual density at any time to the critical density at that time. If $\rho > \rho_c$, then k is positive and the universe is closed; figuratively, one might say that there is then sufficient matter to 'close up the universe'. If $\rho < \rho_c$, the density is too low and the universe is open.

A knowledge of H_0 determines ρ_{c0}, but ρ_0 is more difficult. A different approach to the sign of k is provided by q. From (25.1) and (25.3) we have

$$q = -R\ddot{R}\dot{R}^{-2} = \frac{1}{2}(1 + \frac{3p}{\rho})\frac{\rho}{\rho_c}.$$

For dust this gives

$$q = \frac{1}{2}\frac{\rho}{\rho_c}$$

and for radiation

$$q = \frac{\rho}{\rho_c}.$$

Thus a direct measurement of q_0 (possibly together with a decision between dust and radiation as the main matter content of the universe) will give the sign of k directly. With a knowledge of H_0 and hence ρ_{c0} this will then determine ρ_0 and the single parameter M.

How then do we measure H_0 and q_0? The answer is in the work at the end of chapter 24. Since H and q are essentially the first terms in the Taylor expansion of R as a function of t, we need an observational determination of this function. This comes from knowing the distance of remote objects as a function of their red-shift. As remarked in chapter 24, this in turn requires the adoption of a suitable operational definition of distance and there we defined the apparent area distance d_A. If the universe were conveniently filled with luminous objects of known cross-sectional area at rest relative to the fundamental observers, then we could measure their area distance as a function of their red-shift $d_A(z)$. (In practice, the definition of distance is based on luminosity or *brightness* as in exercise [24.6] and one hopes to find objects of a standard *brightness*, scattered about the universe.) This function is determined theoretically by the following three relations:

(i) definition of z:

$$1 + z = \frac{R(t_0)}{R(t_1)}, \tag{24.13}$$

(ii) definition of d_A:

$$d_A = r_1 R(t_1), \tag{24.16}$$

(iii) equation of the light cone:

$$\int_0^{r_1} \frac{dr}{(1 - kr^2)^{\frac{1}{2}}} = \int_{t_1}^{t_0} \frac{dr}{R(t)}. \tag{24.12}$$

It is then a straightforward calculation to see that the Taylor expansion of $d_A(z)$ begins

$$d_A(z) = \frac{1}{H_0} z + \frac{1}{2} \frac{(q_0 - 3)}{H_0} z^2 + \cdots \tag{25.16}$$

From (25.16) we see at once that for small values of z, distance and red-shift are proportional. This is *Hubble's law*, discovered in 1929 by the American astronomer Edwin Hubble, and the constant of proportionality H_0 is known as *Hubble's constant*. Since all sensible definitions of distance will coincide for small distances, the coefficient of the linear term is independent of the particular definition chosen.

Further, we see that if the relation between distance and red-shift can be determined out to larger red-shifts, then q_0 can be measured and we can establish if the universe is open or closed. Unfortunately there is no *general* consensus on the size of q because of the great difficulties associated with measuring distances.

Exercises for chapter 25

[25.1] Show that in the $k = 1$ universes light leaving the origin at $t = 0$ just has time to go right around and return to the origin in the case of dust, and in the case of radiation only has time to go half way.

[25.2] In 1916, in an attempt to find a static cosmological model, Einstein modified the gravitational field equations by introducing a term proportional to the metric:

$$R_{ab} - \frac{1}{2} R g_{ab} - \lambda g_{ab} = -8\pi G T_{ab}$$

where λ is a constant, the 'cosmological' constant. Show that in the Friedmann-Robertson-Walker models this corresponds to redefining the density and pressure as

$$\tilde{\rho} = \rho + \frac{\lambda}{8\pi G}; \quad \tilde{p} = p - \frac{\lambda}{8\pi G}.$$

Show that, for 'dust' (i.e. $p = 0$, not $\tilde{p} = 0$) the Einstein equations (25.1) and (25.2) do indeed have a static solution and that k must be unity for ρ to be positive. By considering small perurbations in ρ and R show that this static solution is unstable.

[25.3] Write the Friedmann-Robertson-Walker metric in the form:

$$ds^2 = [8\pi G\rho/3 - ke^{2\Omega}]^{-1}d\Omega^2 - [e^{-\Omega}/(1 + kr^2/4)]^2(dx^2 + dy^2 + dz^2)$$

where Ω is a new coordinate. Show that $\rho(\Omega)$ is the density, and that the pressure is governed by the equation

$$\frac{1}{3}\frac{d\rho}{d\Omega} = \rho + p.$$

Hence determine ds^2 for the equations of state $p = (\gamma - 1)\rho$, γ constant; and $p = \omega\rho^{1+1/n}$, ω constant.

[25.4] For $k = -1$ with $p = 0 = \rho$ show that $R(t) = t$. Find a coordinate transformation that reduces this explicitly to flat space-time (cf. exercise 23.4).

[25.5] For a radiation filled universe show that as we approach the big bang ($t = 0$) we have $R \propto t^{1/2}$, irrespective of k. Now suppose the radiation to be *thermal* with temperature T, so that $\rho \propto T^4$. Show that shortly after the big bang $T \propto t^{-1/2}$.

[25.6] Consider the congruence of fundamental world-lines in Friedmann-Robertson-Walker space-time and refer back to exercise 13.2. Find θ and show that both the rotation and the shear vanish. From the shear propagation equation we conclude that the space-time must be free of Weyl curvature. However, the Weyl tensor is conformally invariant (cf. exercise 19.7) so we can further conclude that Friedmann-Robertson-Walker metric must be conformal to flat space; and without the difficult task of finding explicit coordinate transformations.

[25.7] The Hubble constant has dimensions of inverse seconds, but is usually expressed in kilometers per second per megaparsec. (1 $Mpc \sim 3 \times 10^{19} km$.) The value of H_0 is a subject of observational controversy, but as an estimate we may take $H_0 = 50 \ km \ s^{-1}Mpc^{-1}$. Under that assumption show that $\rho_c = 5 \times 10^{-30}g \ cm^{-3}$.

26   Anisotropic cosmologies

ANISOTROPIC COSMOLOGIES constitute a class of cosmological models that are rather more general than the Friedmann-Robertson-Walker models. This class is obtained by dropping the requirement of *isotropy* but retaining *homogeneity*. In particular, this allows these models to have non-zero Weyl tensors, that may in turn change the character of the initial singularities that typically arise.

The definition of homogeneity is by means of an *isometry group*, so we begin with some generalities about *Lie groups* and *Lie algebras*.

As an example, think of the group of 3×3 orthogonal matrices $O(3)$. This is the 3-dimensional rotation group and so is the isometry group of the ordinary round sphere S^2. It is also a manifold since we can think of it as a surface in R^9 defined by the six equations

$$U_{ik}U_{jk} = \delta_{ij} \quad (i, j, k = 1, 2, 3).$$

A group that is also a manifold and for which the group operations are continuous is called a *Lie group*. Here 'continuous' means that a sufficiently small change in two group elements gives a small change in their product, and a sufficiently small change in one element gives a small change in its inverse. Most familiar examples of Lie groups come from matrix groups such as $O(n)$, $U(n)$ and so on.

If a group G acts as a group of transformations (e.g. isometries) on another manifold M, we may consider the corresponding infinitesimal transformations. These define vector fields on M and these vector fields comprise a vector space written g whose dimension is equal to the dimension of G as a manifold. (Think of $G = SO(3)$ acting on the sphere $M = S^2$.) The commutator of two infinitesimal transformations may be defined as the commutator of the two corresponding vector fields on M and this will again be in g. For Killing vectors, i.e. infinitesimal isometries, this was shown in exercise [15.6].

If X, $Y \in g$ we write their commutator as $[X, Y]$ and then find:

$$(a) \ [X, Y] = -[Y, X] \tag{26.1}$$

(b) if $X, Y, Z \in g$ then

$$[X, [Y, Z]] + [Y, [Z, X]] + [Z, [X, Y]] = 0, \tag{26.2}$$

the second of these conditions being known as the *Jacobi identity*. (Cf. exercise 8.7.) A finite-dimensional vector space with an operation [,] that is linear in each component and satisfies (26.1) and (26.2) is called a *Lie algebra*.

Now any Lie group is a manifold and can be made into a group of transformations acting on itself as follows: the element g of G defines the transformation

$$L_g(h) = gh.$$

The *infinitesimal* versions of these transformations therefore define a Lie algebra of special vector fields on G. Since the manifold G has a preferred point, namely the identity e of the group, we can identify this with the tangent space to G at e.

One source of a Lie algebra then is as the tangent space at the identity of a Lie group. There is a converse to this, namely given a Lie algebra it is possible to reconstruct a Lie group from which it arises.

We specify a Lie algebra g by giving a basis $X_1, \ldots, X_n$ and their commutators $[X_i, X_j]$. Since the commutator is again in g, we must have

$$[X_i, X_j] = C_{ij}^k X_k \qquad (26.3)$$

for some set of numbers C_{ij}^k, the *structure constants* of g. From (26.1,2) there are certain conditions on the structure constants which must be satisfied, namely:

$$a) \quad C_{ij}^k = -C_{ji}^k \qquad (26.4)$$

$$b) \quad C_{[ij}^m C_{k]m}^n = 0. \qquad (26.5)$$

Under a change of basis we have

$$X_i \to L_i^j X_j; \quad C_{ij}^k \to L_i^m L_j^n \tilde{L}_p^k C_{mn}^p, \qquad (26.6)$$

where $\tilde{L}_i^j$ is the inverse of L_i^j, so $L_i^j \tilde{L}_j^k = \delta_i^k$.

Our aim in this discussion is to construct homogeneous cosmological models. We want space-times that are foliated by homogeneous space-like hypersurfaces of constant time. Homogeneity for the hypersurfaces will mean that every point is equivalent to every other point, i.e. that there is a group of isometries that is transitive. This in turn requires a 3-dimensional isometry group, so we will make our first task the classification of 3-dimensional Lie algebras. This results in the so-called *Bianchi classification*. Then for each Bianchi *type*, we may write down a 3-dimensional metric with that symmetry and impose the Einstein equations to see what cosmological models result.

Suppose X_i for $i = 1, 2, 3$ forms a basis of a 3-dimensional Lie algebra with structure constants C_{ij}^k. Pick a totally skew tensor ε^{ijk} and define:

$$t^{ij} = \varepsilon^{imn} C_{mn}^j.$$

This contains all the information in C_{ij}^k. We may split it into symmetric and antisymmetric parts and use ε^{ijk} again to write

$$t^{ij} = n^{ij} + \varepsilon^{ijk} a_k$$

$$n^{ij} = n^{(ij)}.$$

The Jacobi identity (26.5) is equivalent to

$$n^{ij} a_j = 0. \tag{26.7}$$

The transformation (26.6), with $det L = 1$ to preserve ε^{ijk}, is

$$n^{ij} \to \tilde{L}^i_p \tilde{L}^j_q n^{pq}, \quad a_j \to L^k_j a_k \tag{26.8}$$

while a new choice of ε^{ijk} has the effect

$$\varepsilon^{ijk} \to \lambda \varepsilon^{ijk}, \quad n^{ij} \to \lambda n^{ij}, \quad a_j \to a_j. \tag{26.9}$$

We must classify a_j and n^{ij} subject to these freedoms. The classification at once splits into two classes:

$$\text{Class } A: \quad a_i = 0,$$

$$\text{Class } B: \quad a_i \neq 0.$$

Dealing first with class A, we may use orthogonal matrices L^j_i in (26.8) to diagonalize n^{ij} and then diagonal matrices L^j_i to set all the non-zero diagonal elements of n^{ij} equal in magnitude. Finally, a suitable choice of λ in (26.9) will set all the non-zero diagonal elements equal to ± 1. This leaves six possibilities, which are set out in table A below with the usual nomenclature.

Table A. The Bianchi types for class A.

n_1	n_2	n_3	Bianchi type
0	0	0	I
1	0	0	II
0	1	-1	VI_0
0	1	1	VII_0
1	1	-1	VIII
1	1	1	IX

In class B, we may use the orthogonal matrices L^j_i in (26.8) to set

$$a_i = \begin{pmatrix} a \\ 0 \\ 0 \end{pmatrix}, \quad n^{ij} = \begin{pmatrix} 0 & 0 & 0 \\ 0 & n_1 & 0 \\ 0 & 0 & n_2 \end{pmatrix}.$$

Now diagonal matrices L^j_i will set n_1 and n_2 equal to ± 1 or 0 and will change a_i in such a way that the quantity

$$h = \frac{a^2}{n_1 n_2}$$

is invariant. Thus some of the class B types have a *continuous* invariant. The six possibilities are given in table B.

Table B. The Bianchi types for class B.

a	n_1	n_2	n_3	Bianchi Type
1	0	0	0	V
1	0	0	1	IV
1	0	1	-1	III or (VI_{-1})
$\sqrt{-h}$	0	1	-1	$VI_h (h < 0)$
$\sqrt{h}$	0	1	1	$VII_h (h > 0)$

This completes the *Bianchi* classification. The different cases are referred to as *Bianchi types*.

Given the classification, we seek for each Bianchi type a 3-manifold with metric and isometry group of that type. This is accomplished with the aid of *invariant one-forms*. Suppose as before that the basis elements X_i of the Lie algebra can be realized as vector fields on a 3-manifold M with the correct commutators. At a point $p \in M$ choose another basis Y_i of vectors with corresponding dual basis ω^i of one-forms (see chapter 11) and propagate these around M by

$$[X_i, Y_j] = 0 \text{ or equivalently } \pounds_{X_i}\omega^j = 0. \tag{26.10}$$

That there is no problem in doing this follows from the Jacobi identity (26.2) with X, Y, Z taken to be X_i, X_j, Y_k. The commutator of any two Y_i can be expressed in terms of the Y_i:

$$[Y_i, Y_j] = \hat{C}^k_{ij} Y_k \tag{26.11}$$

and it follows from the Jacobi identity applied to Y_i, Y_j, X_k that the $\hat{C}^k_{ij}$ are constant on M. For the exterior derivatives of the dual basis we find

$$Y_j^a Y_k^b \nabla_{[a}\omega^i_{b]} = Y_{[j}^a Y_{k]}^b \nabla_a \omega^i_b$$
$$= -\omega^i_b Y_{[j}^a \nabla_{|a|} Y_{k]}^b$$
$$= -\frac{1}{2}\omega^i_b \hat{C}^m_{jk} Y_m^b$$

so

$$d\omega^i = -\frac{1}{2}\hat{C}^i_{jk}\omega^j \wedge \omega^k. \tag{26.12}$$

Finally, if we express the basis Y_i in terms of the basis X_i by

$$Y_i = a_i^j X_j$$

then from (26.10)

$$X_i(a_j^k) = C^m_{ij} a_m^k, \tag{26.13}$$

while from (26.11)

$$\hat{C}^k_{ij} a_k^m = a_i^p a_j^q C^m_{pq}. \tag{26.14}$$

Here (26.13) tells us how to construct the Y_i given the X_i: pick a_i^j at p, then this is a system of differential equations for a_i^j. In particular if we choose $a_i^j = -\delta_i^j$ at p, then

(26.14) says that $\hat{C}^k_{ij}$ is $-C^k_{ij}$ at p. Since the $\hat{C}^k_{ij}$ are constant on M, this will be true everywhere on M; and in particular from (26.12), everywhere on M we will have

$$d\omega^i = \frac{1}{2}C^i_{jk}\omega^j \wedge \omega^k. \tag{26.15}$$

The ω^i are *invariant one-forms* since by (26.10) they are invariant under the group, and the Lie algebra of the X_i is reflected in the formula (26.15). To find suitable coordinates for M for a given type, we solve (26.15) for ω^i given the C^k_{ij} for that type.

Once we have the invariant one-forms for a particular type, any metric of the form

$$d\sigma^2 = a_{ij}\omega^i\omega^j,$$

where a_{ij} is a symmetric matrix of constants, will have the vector fields X_i as Killing vectors. A corresponding *space-time* therefore has metric

$$ds^2 = dt^2 - a_{ij}(t)\omega^i\omega^j \tag{26.16}$$

where t measures proper time along the curves normal to the surfaces of constant t which are homogeneous.

A great deal of work has been done on the various spatially homogeneous cosmologies, and a number of exact solutions are known. We shall concentrate on Bianchi type I models, the simplest case, and note the differences that arise between this case and the Friedmann-Robertson-Walker cosmologies.

From table A we see that for type I the structure constants are all zero. Thus from (26.15) the invariant one-forms are all exact, and we may choose coordinates (x, y, z) so that they are

$$\omega^1 = dx, \quad \omega^2 = dy, \quad \omega^3 = dz.$$

The corresponding Killing vectors are

$$X^1 = -\frac{\partial}{\partial x}, \quad X^2 = -\frac{\partial}{\partial y}, \quad X^3 = -\frac{\partial}{\partial z}$$

so that type I has three translational symmetries (but generally has no rotational symmetries). We may diagonalize the metric $a_{ij}(t)$ at one instant of time and, at least for perfect fluid cosmologies, it will then remain diagonal at all times. The space-time metric is

$$ds^2 = dt^2 - X^2(t)dx^2 - Y^2(t)dy^2 - Z^2(t)dz^2. \tag{26.17}$$

Note that if $X = Y = Z = R$ then this is just the $k = 0$ Friedmann-Robertson-Walker model of chapter 24. This is therefore a generalization that allows for *different rates of expansion in different directions*.

Using the differential form techniques of chapter 11 with the basis

$$\theta^0 = dt, \quad \theta^1 = X dx, \quad \theta^2 = Y dy, \quad \theta^3 = Z dz$$

we find the Ricci tensor to be diagonal with

$$R_{00} = \dot{\theta} + A^2 + B^2 + C^2$$
$$R_{11} = -\dot{A} - \theta A$$
$$R_{22} = -\dot{B} - \theta B$$
$$R_{33} = -\dot{C} - \theta C$$

where

$$A = \frac{\dot{X}}{X}, \quad B = \frac{\dot{Y}}{Y}, \quad C = \frac{\dot{Z}}{Z}, \quad \theta = A + B + C. \tag{26.18}$$

We shall seek perfect fluid solutions with the fluid flow lines parallel to θ^0, i.e. orthogonal to the surfaces of homogeneity. This means

$$R_{00} = -\frac{1}{2}\kappa(\rho + 3p); \quad R_{11} = R_{22} = R_{33} = -\frac{1}{2}\kappa(\rho - p) \tag{26.19}$$

in terms of density ρ, pressure p, and $\kappa = 8\pi G/c^2$. The conservation equation (24.6) is

$$\dot{\rho} + \theta(\rho + p) = 0. \tag{26.20}$$

We shall look first for *vacuum* solutions. From (26.18) and (26.19) we find

$$2R_{00} - R = -2(AB + BC + CA) = -2\kappa\rho,$$

which is zero for vacuum, so that

$$\theta^2 = A^2 + B^2 + C^2. \tag{26.21}$$

Now $R_{00} = \dot{\theta} + \theta^2 = 0$ so that

$$\theta = \frac{1}{t - t_0}. \tag{26.22}$$

By choice of origin in t we may set t_0 to zero. Then

$$R_{11} = -\dot{A} - A\theta = 0 \Longrightarrow A = pt^{-1} \quad (p \text{ constant}).$$

and similarly $B = qt^{-1}$, $C = rt^{-1}$. Next (26.22) and (26.21) imply

$$p + q + r = 1 \tag{26.23a}$$

$$p^2 + q^2 + r^2 = 1 \tag{26.23b}$$

and finally

$$A = \frac{\dot{X}}{X} = \frac{p}{t} \Longrightarrow X = X_0 t^p.$$

The metric is therefore

$$ds^2 = dt^2 - t^{2p}dx^2 - t^{2q}dy^2 - t^{2r}dz^2 \tag{26.24}$$

with p, q, r subject to (26.23), where we have absorbed some inessential constants of integration. This is known as the *Kasner metric*. We shall discuss its properties after finding the corresponding *dust* cosmology. For the dust cosmology $p = 0$ so from (26.18,19) we have

$$R_{11} + R_{22} + R_{33} = -\dot{\theta} - \theta^2 = -\frac{3}{2}\kappa\rho.$$

Therefore, by use of the conservation equation, we get

$$(\dot{\theta} + \theta^2)^{\bullet} = \frac{3}{2}\kappa\theta\rho = -\frac{3}{2}\kappa\theta\rho = -\theta(\dot{\theta} + \theta^2).$$

To solve this, set $\theta = \dot{v}/v$, then

$$\left(\frac{\ddot{v}}{v}\right)^{\bullet} = -\frac{\dot{v}}{v}\frac{\ddot{v}}{v},$$

i.e.

$$\dddot{v} = 0 \quad \text{and} \quad v = \alpha t^2 + 2\beta t + \gamma,$$

so

$$\theta = \frac{2(\alpha t + \beta)}{\alpha t^2 + 2\beta t + \gamma}, \quad \kappa\rho = \frac{4\alpha}{3v}.$$

We must have $\alpha \neq 0$, so without loss of generality suppose $\alpha = 1$, and by choice of origin set $\beta = 0$:

$$\theta = \frac{2t}{t^2 + \gamma}, \quad \kappa\rho = \frac{4}{3(t^2 + \gamma)} = \frac{4}{3v}. \tag{26.25}$$

From (26.18) and (26.19) we have

$$-R_{11} = \dot{A} + \theta A = \frac{2}{3v} \implies (vA)^{\cdot} = \frac{2}{3}$$

and

$$A = \frac{2}{3}\frac{t + p_0}{t^2 + \gamma}$$

for constant p_0. Similarly,

$$B = \frac{2}{3}\frac{t + q_0}{t^2 + \gamma}, \quad C = \frac{2}{3}\frac{t + r_0}{t^2 + \gamma}.$$

From (26.18) again

$$A + B + C = \theta \implies p_0 + q_0 + r_0 = 0$$

$$AB + BC + CA = \kappa\rho \implies p_0 q_0 + q_0 r_0 + r_0 p_0 = 3\gamma.$$

Note that:

$$2(p_0 q_0 + q_0 r_0 + r_0 p_0) = (p_0 + q_0 + r_0)^2 - p_0^2 - q_0^2 - r_0^2 < 0$$

so

$$\gamma < 0, \quad \text{say } \gamma = -k^2.$$

Finally for the metric components, absorbing inessential constants of integration, we get

$$\frac{\dot{X}}{X} = A \implies X = (t - k)^p (t + k)^{2/3 - p}, \tag{26.26}$$

and similarly

$$Y = (t - k)^q (t + k)^{2/3-q}$$
$$Z = (t - k)^r (t + k)^{2/3-r}.$$

with

$$p = \frac{1}{3}(1 + \frac{p_0}{k}).$$

and so on, so that

$$p + q + r = 1 \tag{26.27a}$$
$$p^2 + q^2 + r^2 = 1. \tag{26.27b}$$

This gives the *dust cosmology* counterpart to the Kasner metric (26.24). Notice that for large t it tends to the dust Friedmann-Robertson-Walker cosmology (25.7), and for zero k it is precisely the Friedmann-Robertson-Walker model.

There will be a singularity of the metric at $t = k$, and from (26.25) since the density is given by

$$\kappa\rho = \frac{4}{3(t^2 - k^2)}$$

this is genuinely a curvature singularity. However for non-zero k, ρ has a simple pole, rather than a double pole as it would be here for the Robertson-Walker model. Near the singularity, the form of (26.26) approaches the Kasner metric. To analyse the singularity, we need to solve (26.23). In (p, q, r)-space these are the equations of a plane and a sphere which therefore intersect in a circle in fact through the points $(1, 0, 0)$, $(0, 1, 0)$, $(0, 0, 1)$ (see figure 26.1). We may choose the solution so that

$$-\frac{1}{3} < p < 0 < q < r < 1$$

and parametrize it by u with $0 < u < 1$ according to

$$p = \frac{-u}{1 + u + u^2}, \quad q = \frac{u(u + 1)}{1 + u + u^2}, \quad r = \frac{1 + u}{1 + u + u^2}.$$

Necessarily then one of the exponents in (26.24) and (26.26) is negative and two are positive unless two of them vanish (a case which, at least for the vacuum metric we rule out, see exercise 26.7). This means that as we approach the singularity two metric components go to zero while the third goes to infinity. Thus, a small comoving volume of fluid is squashed to zero in two directions and stretched to infinity in the third, while its volume shrinks to zero. This is to be contrasted with the Friedmann-Robertson-Walker singularity where the volume shrinks uniformly in every direction. The difference is due to the Weyl tensor, which diverges at the singularity. In the Kasner solution it is clear that this happens, since there is no Ricci tensor, so any singularity *must* be in the Weyl tensor and there certainly is a singularity (as may be seen from the curvature invariant $C^{abcd}C_{abcd}$ which turns out to be $t^{-4}u^2 f(u)$ where f is a strictly positive polynomial in u). In the dust cosmology it is clear from the way the metric approaches the Kasner metric that here too the Weyl tensor diverges.

Many other examples are known of spatially homogeneous cosmologies of different Bianchi types. Some have, like this one, fluid matter content flowing orthogonally to the surfaces of homogeneity, others have fluid *not* orthogonal to these surfaces, and it is also possible to include electromagnetic fields or mixtures of fluids. Most have Weyl tensor singularities, and in some cases these can have a very complicated structure. They provide a useful menagerie of examples generalizing the Friedmann-Robertson-Walker models and providing experience in building relativistic cosmological models. They are not regarded in themselves very seriously as *realistic* cosmological models since the observational evidence for a significant degree of isotropy of the universe is so strong. However, each of the Friedmann-Robertson-Walker cosmologies is a particular case of a spatially homogeneous, anisotropic cosmology, and one question that must be answered is why the actual universe is so much *less* anisotropic than the equations in principle allow it to be. One might even go so far as to speculate, as Roger Penrose has done, that the 'initial conditions' on the structure of the universe in some way logically entail isotropy—or, alternatively, a vanishing Weyl tensor.

Exercises for chapter 26

[26.1] Show that the Lie algebra formed by the Killing vectors K_2, K_3 and K_4 of the Schwarzschild solution (exercise 17.3) form a Lie algebra of Bianchi type IX.

[26.2] A basis of one-forms on a 3-manifold M is defined by:

$$\omega^1 = \cos\psi d\theta + \sin\psi\sin\theta d\phi$$
$$\omega^2 = \sin\psi d\theta - \cos\psi\sin\theta d\phi$$
$$\omega^3 = d\psi + \cos\theta d\phi$$

in terms of angular coordinates θ, ϕ, ψ. For what C^k_{ij} do these one-forms satisfy (26.15)? What is the Bianchi type of the metric

$$A(\omega^1)^2 + B(\omega^2)^2 + C(\omega^3)^2$$

and what are its Killing vectors?

[26.3] Find the curvature of the metric in exercise 26.2 using the Cartan calculus. What happens if $A = B = C$? What is the Bianchi type of the $k = 1$ Friedmann-Robertson-Walker model?

[26.4] Repeat exercises 26.2 and 26.3 with the basis

$$\omega^1 = dx, \quad \omega^2 = e^x dy, \quad \omega^3 = e^x dz$$

This should tell you the Bianchi type of the $k = -1$ Friedmann-Robertson-Walker model.

[26.5] We consider the group $0(1,2)$ defined as the 3×3 real matrices M_{ij} satisfying

$$\eta_{jk} M_{ji} M_{kl} = \eta_{il}, \quad \text{where } \eta_{ij} = diag(1,-1,-1).$$

To find a way of writing the Lie algebra, consider an element of $0(1,2)$ near to the identity: $M_{ij} = \delta_{ij} + \epsilon K_{ij}$. Show that to first order in ϵ

$$\eta_{lj} K_{ji} + \eta_{ik} K_{kl} = 0$$

so that $\eta_{ij} K_{jk} = A_{ik}$ is skew and $K_{ik} = \eta_{ij} A_{jk}$. As a basis for such K_{ij} consider the set

$$X_1 = \begin{pmatrix} 0 & 1 & 0 \\ 1 & 0 & 0 \\ 0 & 0 & 0 \end{pmatrix}, \quad X_2 = \begin{pmatrix} 0 & 0 & 1 \\ 0 & 0 & 0 \\ 1 & 0 & 0 \end{pmatrix}, \quad X_3 = \begin{pmatrix} 0 & 0 & 0 \\ 0 & 0 & -1 \\ 0 & 1 & 0 \end{pmatrix}.$$

What are the structure constants of the Lie algebra with respect to this basis if the Lie bracket is just the matrix commutator? What Bianchi type is this? This example indicates a general approach for dealing with matrix groups.

[26.7] Show that the Kasner metric (26.24) with $p = q = 0$, $r = 1$ is flat. (The easiest way is to guess the coordinate transformation that makes it manifestly flat. You need 'hyperbolic' polar coordinates involving t and z).

[26.8] Show that if a Bianchi type I cosmology has an ideal fluid content then the fluid flow lines are *necessarily* orthogonal to the hypersurfaces of homogeneity.

[26.9] With the introduction of a new time coordinate Ω show that the general Bianchi type I cosmology with an ideal fluid content can be expressed as

$$ds^2 = \Phi^{-1} d\Omega^2 - e^{-2\Omega} \sum_{i=1}^{3} (e^{B_i \Psi} dx^i)^2$$

where

$$\Psi = \int \Phi^{-1} e^{3\Omega} d\Omega$$

and

$$\Phi = \frac{8\pi G \rho}{3} + \frac{1}{6} B^2 e^{6\Omega}.$$

The constants B_i satisfy $\sum B_i = 0$, and B^2 is given by $\sum B_i^2$. The pressure $p(\Omega)$ is determined by

$$p = \frac{1}{3} \frac{d\rho}{d\Omega} - \rho.$$

Cf. exercise [25.3].

Index

Printed in the United States
By Bookmasters